# Strategic Plan

## FY 2010-2015

 **United States Department of Agriculture**

# Table of Contents

# Message from
# the Secretary

Now more than ever, America's agricultural and rural communities face challenges that jeopardize the livelihoods and well-being of people working the land or living in rural areas. As this strategic plan was being written, the Nation was working to pull itself out of the worst recession since the Great Depression. These economic circumstances have led to unpredictable input costs and unsteady demands for agricultural products in a rapidly evolving global marketplace. More Americans are hungry than at any time in the past 15 years, and our children are increasingly at risk of growing up overweight or obese. In the longer term, a changing global climate brings increased uncertainties about the effect of weather patterns on crop production and the conservation of our natural resources, and puts a premium on improving energy efficiency and producing a renewable energy supply at home. The Department is committed to a strong safety net for production of agriculture.

These challenges and others will require USDA to not only provide a reliable safety net for farmers and ranchers, but also help communities and businesses to innovate by implementing new technologies and modernizing their infrastructure to ensure access to new markets, increased competitiveness, and greater resilience of their productive resources.

At the same time, these challenges create many opportunities for farmers, ranchers, forest landowners, public land managers, and families in rural communities to generate prosperity in new ways while conserving the Nation's natural resources and providing a safe, sufficient, and nutritious food supply for the country and the world.

Because USDA programs touch almost every American every day, the Department is well positioned to support its constituents in taking advantage of these new opportunities. To ensure the Department's programs deliver results effectively and efficiently, USDA's Strategic Plan for fiscal years (FY) 2010-2015 lays out key policy priorities and the strategies to achieve them. Over the next 5 years, USDA will use this plan to manage its resources in a way that delivers the best outcomes for everyone affected by its diverse program portfolio.

The USDA Strategic Plan for FY 2010-2015 differs from previous plans by striving to break down organizational barriers to maximize the Department's available resources. Key priorities and desired outcomes have been identified, as well as the best means and strategies to achieve them. In the pursuit of these outcomes, agencies and offices of USDA will be encouraged to collaborate more effectively to achieve the shared goals of rural prosperity, preservation and maintenance of forests and working lands, sustainable agricultural production, global food security, and safe and nutritious foods for Americans.

This strategic plan represents the dynamic process within USDA to ensure the best results for America. Through this process, the Department is able to continually assess the quality of its provision of services to the public. This close attention to performance outcomes and results will allow USDA to better support its constituents as they strive to take advantage of today's new opportunities.

*Thomas J. Vilsack*

Thomas J. Vilsack
Secretary of Agriculture

## Mission Statement

We provide leadership on food, agriculture, natural resources, rural development, nutrition, and related issues based on sound public policy, the best available science, and efficient management.

## Vision Statement

To expand economic opportunity through innovation, helping rural America to thrive; to promote agriculture production sustainability that better nourishes Americans while also helping feed others throughout the world; and to preserve and conserve our Nation's natural resources through restored forests, improved watersheds, and healthy private working lands.

## Core Values

Our success depends on:

- Transparency — Making the Department's management processes more open so that the public can learn how USDA supports Americans every day in every way.

- Participation — Providing opportunities for USDA constituents to shape and improve services provided by the Department.

- Collaboration — Working cooperatively at all governmental levels domestically and internationally on policy matters affecting a broad audience.

- Accountability — Ensuring that the performance of all employees is measured against the achievement of the Department's strategic goals.

- Customer Focus — Serving USDA's constituents by delivering programs that address their diverse needs.

- Professionalism — Building and maintaining a highly skilled, diverse, and compassionate workforce.

- Results Orientation — Measuring performance and making management decisions to direct resources to where they are used most effectively.

# Strategic Plan Framework

Founded by President Abraham Lincoln in 1862, when more than half of the Nation's population lived and worked on farms, USDA's role has evolved with the economy. Today, the country looks to rural America to not only provide food and fiber, but also for crucial emerging economic opportunities in renewable energy, broadband, and recreation. People in rural areas operate in a technologically advanced, rapidly diversifying, and highly competitive business environment driven by increasingly sophisticated consumers.

To assist the country in addressing today's challenges, USDA will:

- Assist rural communities to create prosperity so they are self-sustaining, re-populating, and economically thriving (Goal 1);

- Ensure our national forests and private working lands are conserved, restored, and made more resilient to climate change, while enhancing our water resources (Goal 2);

- Help America promote agricultural production and biotechnology exports as America works to increase food security (Goal 3); and

- Ensure that all of America's children have access to safe, nutritious, and balanced meals (Goal 4).

These four strategic goals articulate the Department's priorities. These goals contain 14 objectives that describe the Department's major programmatic policies and cover the myriad programs and services that USDA administers. USDA's strategic goals mirror the Department's commitment to provide exceptional service and state-of-the-art science through consistent management excellence across the Department.

Sound management is an inherent part of achieving these goals. The Department has separate management plans that detail USDA's strategies to achieve its goals in the areas of human capital, outreach efforts, performance and efficiency, information technology, "green" operations, homeland security, and emergency preparedness.

In addition, more detailed mission area and agency plans are being developed to support this strategic plan.

Performance measures will track progress in attaining each objective and its overarching goal. Measures specify baseline information and long-term performance targets. Strategies and means describe the actions that need to be taken to accomplish the strategic goals. The external risk factors section of each goal highlights possible challenges the USDA may encounter in making progress on each strategic goal.

 is the leading advocate for rural America. The Department supports rural communities and enhances quality of life for rural residents by improving their economic opportunities, community infrastructure, environmental health, and the sustainability of agricultural production. The common goal is to help create thriving rural communities where people want to live and raise families, and where the children have economic opportunities and a bright future.

USDA revitalizes rural communities by expanding economic opportunities and creating jobs for rural residents. USDA, in cooperation with its public and private partners, is connecting rural residents to the global economy by expanding access to broadband to unserved and underserved communities; promoting rural leadership in sustainable renewable energy development; creating new opportunities for small agricultural producers to market their products by developing local and regional food systems; ensuring that rural residents capitalize on potential opportunities presented by the Nation's efforts to develop markets for ecosystem services and mitigate climate change; and generating jobs through recreation and natural resource conservation, restoration, and management in rural areas. USDA operates job training and business development programs that give rural residents the tools and capacity to access markets and enter the green economy.

USDA is working to enhance the livability of rural communities. The Department uses 21st century technology to rebuild infrastructure, ensure that rural residents have decent housing and homeownership opportunities, clean water, adequate systems for handling waste, reliable electricity and renewable energy systems, and critical community facilities including health-care centers, schools, and public safety departments. USDA also helps communities invest in strategic green-infrastructure planning and protection of critical natural resources.

The economic vitality and quality of life in rural America also depends on a financially healthy agricultural system and access to agricultural markets. The country's farmers help ensure that all of America and many other parts of the world have nutritious and safe food, adequate energy sources, and fiber products sufficient to meet the needs of our rapidly growing population. USDA works to ensure American farmers and ranchers are competitive and producers have access to new and international markets, adequate support in times of economic or environmental distress, and the ability to manage their risks. The Department strives to provide agricultural producers with an adequate safety net comprised of necessary risk management tools, disaster assistance, and prompt and equitable assistance for farmers, ranchers, and eligible landowners. USDA encourages producers to be good stewards of their lands so American agricultural production is economically and environmentally sustainable, as well as socially beneficial.

USDA will achieve this goal through a focus on asset and data-driven investment decisions coupled with strategic place-based decision making. The Department will provide on-the-ground support (financial, technical, and planning assistance) for local multi-county, community-driven strategic plans. USDA will also use the Rural Innovation Initiative to promote economic opportunity and job creation in rural communities. These investments will facilitate and support regional economic development by combining a multitude of financial and technical resources to maximize the collaborative economic development impact on high-priority regions. As part of this strategy, USDA will create partnerships to leverage investments made by other Federal departments, tribal, State, and local partners, and private entities to more effectively support rural communities and regions. These investments allow and support our long-term national prosperity by ensuring that rural communities are self-sustaining, repopulating, and thriving economically.

Over the next 5 years, USDA will work to enhance rural prosperity (Objective 1.1), create livable communities (Objective 1.2), and support a sustainable, competitive agricultural system (Objective 1.3).

## OBJECTIVE 1.1 – ENHANCE RURAL PROSPERITY

While rural communities face unique challenges in creating and sustaining good jobs, they are also presented with unprecedented opportunities for economic growth. The Department works to ensure that rural residents have the ability to capture these opportunities so that rural families thrive and rural youth see a bright future in their hometowns.

USDA revitalizes rural communities by creating jobs[1] and expanding economic opportunities for rural residents. (See performance measure 1.1.1.) The Department has identified five pillars that support the achievement of this objective: (1) increasing access to broadband; (2) facilitating sustainable renewable energy development; (3) developing regional food systems; (4) capitalizing on climate change opportunities; and (5) generating and retaining jobs through recreation and natural resource restoration, conservation, and management. In addition, USDA operates business development and job training programs to give citizens advanced community development opportunities, continuous business creation opportunities, and the tools and capabilities to access markets and gain employment.

### Pillar 1 — Increasing Access to Broadband

Expanding broadband capacity into communities that otherwise might not have access provides businesses and communities increased opportunity to form networks and connect to the global economy. USDA will deploy broadband infrastructure in unserved and underserved areas in the United States by targeting outreach, expanding the availability of public computer centers, and encouraging the adoption of broadband service. Increased access to broadband will help rural communities attract new business and cooperative development; increase local leadership development; and improve community services and capacity through community connect, distance learning, and telemedicine programs. (See performance measure 1.1.2.)

---

[1] With a focus on quality job creation; quality is defined as jobs that pay wages that average at least 125 percent of the Federal minimum wage; qualify under the Work Opportunity Tax Credit Program authorized by the Small Business and Work Opportunity Tax Act of 2007; or offer a health care benefits package to all employees, with at least 50 percent of the premium paid by the employer.

### Pillar 2 — Facilitate Sustainable Renewable Energy Development

USDA promotes rural America's role in renewable energy production by providing leadership in the research, development, and sustainability of renewable energy and energy efficiency. The Department's activities will support the reduction of both America's greenhouse gas emissions and dependence on foreign oil. USDA programs make it economically feasible for farmers, ranchers, and other rural small businesses to invest in alternative energy production and energy-saving activities. Through funding opportunities in the form of payments, grants, loans, and loan guarantees for energy projects such as biomass, biofuels, wind, solar, geothermal, hydro-electric, and ocean waves, the Department is committed to developing clean energy sources, promoting energy efficiency, and curbing the effects of climate change. When compatible with other natural resource goals, USDA will facilitate the use of public lands in our National Forest System to support sustainable renewable energy development. The Department will play a role in developing and deploying the sustainable use of biofuels as a renewable energy resource. The future of the biofuels industry must also include the commercialization of second- and third-generation feedstocks and the use of non-food sources. USDA programs can also target financial assistance for gains in energy efficiency. Renewable energy loans and grants foster energy independence as farmers, ranchers, and other rural businesses convert from expensive and environmentally unsustainable, traditional energy consumption to alternative sources. The distributed nature of these renewable resources holds great potential for increased employment, prosperity, and local energy in rural areas. (See performance measure 1.1.3.)

### Pillar 3 — Develop and Support Regional Food Systems

An increased emphasis on regional food systems will have direct and significant benefits to rural communities. Increased economic activity in food-related sectors of the economy help communities build and maintain prosperity. Building on the foundation established in the 2008 Farm Bill, USDA will work closely with all its strategic partners – Federal, tribal, State, county, local, community, and private sector – to develop and revitalize the critical infrastructure necessary for vibrant regional food systems. This includes supporting innovative new opportunities as well as proven business approaches like cooperatives. Farmers markets continue to provide an important distribution channel for direct-to-consumer sales. Regional food hubs will also play a role in enabling regional food systems to serve traditional and institutional customers.

USDA continues to support the planning, coordination, and education necessary for thriving regional food systems, and the recognition and replication of successful models. (See performance measure 1.1.6.)

## Pillar 4 — Capitalize on Opportunities Presented by the Nation's Efforts to Develop Markets for Ecosystem Services and Mitigate Climate Change

Climate change is a critical challenge facing the United States and the world. Farmers, ranchers, and forest landowners have a role to play in addressing climate change. The Nation's response to climate change also represents a potential opportunity to create wealth and revitalize communities in rural America. Emissions from domestic agriculture account for 6 percent of overall U.S. greenhouse gas emissions, while U.S. lands sequester approximately 12 percent of the Nation's annual greenhouse gas emissions. By expanding stewardship practices such as conservation tillage, afforestation, construction of methane digesters, improved forest management, and nutrient use efficiency, landowners could play a role in reducing greenhouse gas emissions and increasing carbon sequestration. The President has called for an economy-wide target of reducing emissions in 2020 to 17 percent below 2005 levels. USDA conservation and energy programs will target actions to reduce greenhouse gas emissions and increase carbon sequestration.

The potential emergence of a viable greenhouse gas offsets market—one that rewards farmers, ranchers, and forest landowners for permanent stewardship activities—also has the potential to play a role in helping America become energy independent and in the Nation's efforts to reduce our greenhouse gas emissions. Further, the potential for new environmental markets for other ecosystem services, such as improved water quality, water conservation, and biodiversity, could provide landowners with significant new sources of revenue.

Climate change is already affecting U.S. agriculture, land and water resources, and biodiversity, and will continue to do so, creating risks that threaten rural prosperity. USDA has a responsibility to support efforts of landowners in adapting to climate change through conservation practices that conserve water, reduce catastrophic fires, and protect and restore flood-prone areas, among others. USDA will expand its work with landowners to increase adoption of practices that will make farms, ranches, and forestlands more resilient to the effects of climate change. (See performance measure 1.1.4.)

## Pillar 5 — Generate and Retain Green Jobs and Economic Benefits Through Natural Resource and Recreation Programs

USDA is a key player in the President's America's Great Outdoors initiative to build a 21st century conservation agenda, based on the experience and ideas of people from across the country. The goal of the initiative is to conserve our natural resources, both public and private, while reconnecting Americans to the outdoors.

USDA natural resources conservation, restoration, and land management programs generate and sustain rural jobs and prosperity. Forests and grasslands generate economic value by attracting tourism and recreation visitors; sustaining green jobs; and producing timber and non-timber products, minerals, food, water, and renewable and non-renewable energy. USDA recreation programs create direct and indirect recreation, tourism, and related service-sector jobs and economic benefits. These benefits include work restoring or creating trails and facilities; providing camping, boating, and outfitter and guide services; and supporting associated retail businesses. Hunting and fishing activity is also an important economic driver in rural America, and leads people to protect habitat and maintain interest in the outdoors.

Other programs generate restoration jobs, which reduce hazardous fuel loads, restore wildlife and fish habitats, and improve watershed conditions on national forests and grasslands, and tribal, State, and private lands. USDA also works with farmers, ranchers, and others to conserve and restore private wetlands, croplands, wildlife habitats, and riparian areas. The Department's partnerships with farmers, ranchers, and communities create corridors for wildlife, prevent and reduce impacts from flooding or other extreme weather events, and support the creation of green space for rural recreational use. (See performance measure 1.1.5.)

To supplement these five pillars and further support rural regional economic prosperity, USDA, both by itself and by working with partners, provides job training and business development opportunities for rural and urban residents. These include the natural resources-based Job Corps and Youth Conservation Corps for young adults; cooperative business development; community economic development and strategic community planning; along with faith-based and self-help initiatives. In 2009, the USDA graduated 2,779 students from USDA-run Job Corps Centers. These and other programs provide the educational opportunities, training, technical support, and tools for rural residents to start small businesses and access jobs in agricultural markets, the green economy, and other existing markets, as well as acquire training in vocational and entrepreneurship skills they can use in the marketplace and business sector.

## Performance Measures

1.1.1 Annual number of jobs created or saved through investments in business, entrepreneurship, cooperatives, and industry

| Baseline 2009 | | Target 2015 |
|---|---|---|
| 68,969 | | 85,000 |

| 2006 | 2007 | 2008 |
|---|---|---|
| 71,715 | 72,710 | 72,907 |

1.1.2 Annual number of subscribers receiving new or improved broadband telecommunication and services

| Baseline 2009 | | Target 2015 |
|---|---|---|
| 190,000 | | 402,500 |

| 2006 | 2007 | 2008 |
|---|---|---|
| 300,000 | 360,000 | 780,000[2] |

1.1.3 Millions of kilowatt hours (mkWh) generated in rural America from alternative energy sources

| Baseline 2009 | | Target 2015 |
|---|---|---|
| 1,504 mkWh | | 3,123 mkWh |

1.14 Annual revenue generated from environmental markets including agricultural and forestry-based carbon offsets, wetland banking, conservation banking, and water quality credit trading

| Baseline 2008 | Target 2015 |
|---|---|
| $1.9 billion[3] | $4.0 billion |

1.1.5 Annual economic contribution of recreation on National Forests and Grasslands

| Baseline 2008 | Target 2015 |
|---|---|
| 237,800 jobs | 247,300 jobs |

1.1.6 Cumulative number of farmers markets established, increasing consumer access to local food

| Baseline 2009 | Target 2015 |
|---|---|
| 5,274 | 6,300 |
| ($1.22 billion in sales per year) | ($1.5 billion in sales per year) |

## Strategies and Means

USDA will create opportunities in rural communities by encouraging communities to come together to form regional strategies. Through grassroots, asset-based strategies that rely on strong analytics, this approach will empower rural citizens and ensure that Federal dollars are better targeted. The approach will encourage collaboration among Federal, State, and local governments, as well as between the government, private sector, non-profit community, and educational institutions. USDA resources will also be used to initiate regional planning efforts with tribes. They will also be used with other Departmental assistive efforts, such as planning grants, leveraging of resources, additional funding, access to credit, and similar efforts to address the unique challenges in

---

[2] One broadband loan for $267 million accounted for 447,114 subscribers, which was an anomaly for FY 2008.

[3] 2010 State of the Biodiversity Markets (ecosystemmarket place.com) includes a range of $1.1 -1.8 billion per year for wetland banking and $200 million per year for conservation banking. 2010 State of the Water Quality Trading Market prerelease data (also from ecosystemmarketplace.com) includes $10.8 million for water quality trading. Point Carbon's annual survey of carbon market transactions shows the US 2008 carbon trading market to be a total of $65 million. Agriculture- and forestry-related transactions constituted about a third of those transactions, or conservatively $32 million.

rural economic development that are part of the experience of tribal governments and communities.

Through multiple agencies, USDA will apply several broader strategies to enhance rural prosperity. These include leveraging technology and innovation, encouraging business development, regional planning, and increasing available funding and access to credit.

Leveraging technology and innovation would entail using technology and innovation in distributing business tools, information, and resources. It would also use non-formal education programs, outreach, 4-H, and other youth development programs to transfer knowledge and technology.

Encouraging business development will involve providing up-to-date training for field staff to properly utilize these five pillars in directing activities and investments. It would also entail identifying and marketing business programs in high-need areas. The Rural Innovation Initiative is designed to promote economic opportunity and job creation in rural communities. To support this innovative approach, USDA plans to set-aside more than $130 million, or roughly 5 percent of the funding from approximately 20 existing programs. These funds would be allocated competitively among regional pilot projects tailored to local needs and opportunities. This targeting effort will allow the Department to prioritize areas with the greatest need and potential by encouraging comprehensive and innovative approaches to foster rural revitalization.

Assisting rural communities in meeting their funding needs as well as increasing their access to credit means applying strategies that include maintaining a farm credit system with sound collateral, loan, and guaranteed-loan portfolio oversight to secure investments. It also means participating in marketing efforts to reach national lenders (capital markets) and leveraging USDA resources with private capital. The Department seeks to provide equitable access to capital for rural minority business owners and targeted outreach to underserved and minority communities. It also provides on-the-ground technical assistance to communities for access and resources to Federal investments. USDA seeks to enhance rural regions by supporting community and multi-county driven strategic plans that connect regions to emerging and existing industry sectors.

Specific strategies and means applied to the five pillars include:

Pillar 1 — Increasing access to broadband:

- Perform outreach activities to inform rural communities of the benefits of broadband; and
- Provide loans and grants, and loan and grant combinations to telecommunications firms and cooperative service organizations to support broadband deployment within unserved and underserved communities.

Pillar 2 - Facilitate sustainable renewable energy development:

- Support expanding production of advanced biofuels through the financing of the widespread deployment of full-scale commercial facilities;
- Evaluate programs and coordinate efforts across USDA and other Federal departments and agencies to support production capacity, sustainable commercialization, and distribution of biofuels to accomplish the Renewable Fuel Standards mandate (RFS2) while establishing markets and demand for biofuels;
- Create viable engineering, business, and financial protocols to evaluate proposed commercial renewable energy and energy efficiency projects to benefit both the commercial and government sectors;
- Develop programs to help industry utilize new science and technology in sustainably producing and transforming the Nation's renewable biomass, wind, solar, and geothermal resources into cost competitive renewable energy;
- Integrate existing research and development areas for bioenergy and climate change, food and fuel, and renewable energy production/use and the environment;
- Support the establishment and production of eligible crops such as annual and non-woody perennial crops for conversion to bioenergy by assisting agricultural and forest land owners and operators with the collection, harvest, storage, and transportation of renewable biomass for conversion of bioenergy;
- Integrate renewable energy production into sustainable agriculture, forest, and range management systems;
- Develop superior genetic biofuel feedstocks and the needed sustainable production and logistic systems

suited to regional conditions and biofuel refinery specifications; and

- Develop processes and technologies to produce value-added bioproducts that can be added alongside biofuels production to diversify product options, increase income, and diversify risks.

Pillar 3 — Develop and support regional food systems:

- Connect investments to regional economic development strategies, such as food hubs;
- Use job training and business development programs across urban and rural communities to reduce the rural-urban divide;
- Provide regions with an economic analysis to better support decision making;
- Develop local leadership and increase productive capacity for processing, storage, and distribution;
- Develop and support educational outreach to regions to assess local food systems; and
- Connect regional food systems to existing markets (e.g., universities, community facilities, school programs).

Pillar 4 — Capitalize on opportunities presented by the Nation's efforts to develop markets for ecosystem services and mitigate climate change:

- Develop and assist entry to markets for greenhouse gas offsets to help landowners benefit from opportunities presented by climate change;
- Develop and facilitate participation in multiple ecosystem markets to help landowners generate new income by capturing the economic benefits from ecosystem services
- Provide technical and financial assistance for conservation, renewable energy, and energy efficiency actions to reduce greenhouse gas emissions;
- Support innovative strategies to capitalize on climate change mitigation;
- Provide technical support and analysis to enhance water conservation and restore watershed health; and
- Increase research and development efforts to support water conservation.

Pillar 5 — Generate and retain green jobs and economic benefits through natural resource and recreation programs:

- Provide recreational opportunities in National Forests that offer healthy activity and generate revenue from tourism;
- Create jobs for youth in rural America by funding projects that improve or maintain recreation facilities, particularly trails, on public lands;
- Provide more opportunities for natural resource conservation and restoration work on public lands;
- Support the production of non-timber products, wood, and energy where consistent with natural resource goals;
- Provide technical and financial assistance for conservation work on private lands;
- Collaboratively engage public lands communities in natural resource management;
- Support pathways for entry to the green economy by providing training to youth and adults through job corps centers and youth corps programs;
- Connect investments to regional economic development strategies;
- Participate in and organize public listening sessions for America's Great Outdoors;
- Work with stakeholders to develop a conservation agenda and connect people to the outdoors; and
- Facilitate the protection of habitat (through restoration and management) and access for hunting and fishing on public and private lands.

## OBJECTIVE 1.2 – CREATE THRIVING COMMUNITIES

USDA seeks to enhance the opportunities necessary for rural families to thrive economically and to increase the quality of life in rural communities so that these communities are places where people want to live. USDA uses 21st century technology to rebuild the infrastructure of small communities, ensuring that rural residents have decent housing and homeownership opportunities, clean water, adequate systems for handling waste, reliable electricity and renewable energy systems, and vital community facilities, including critical health-care centers, schools, faith-based initiatives, and public safety departments. Affordable housing, both homeownership and rental housing, is a critical component to rural economic vitality. Investments in residential infrastructure, such as water and waste facilities, schools, and public safety departments, enable production of new and revitalized housing needed in thriving rural communities. These rural development programs enhance rural prosperity by establishing a foundation for regional and community economic growth while protecting the natural resources people value.

The Department also works to ensure that rural residents live in a healthy and productive environment, with clean air, clean water, and access to outdoor recreation opportunities. A healthy environment is an essential ingredient for thriving, sustainable rural communities and rural quality of life. In addition to enhancing the livability of communities, strategic green infrastructure planning and strategic place-based efforts can provide energy conservation, stormwater pollution abatement, improved public physical and mental health, and other values. USDA partners with communities, local businesses and foundations, and many others to establish urban and community forestry programs. These partnerships also create green infrastructure to maintain, restore, and enhance the natural resources and quality of life in urban and rural communities.

## Performance Measures

1.2.1 Annual number of subscribers receiving new or improved water and waste facilities services

| Baseline 2007 | | Target 2015 | |
|---|---|---|---|
| 1,322,063 | | 2,000,000 | |

| 2006 | 2008 | 2009 |
|---|---|---|
| 1,637,554 | 4,361,972[4] | 3,400,000[4] |

1.2.2 Annual number of (1) homeownership opportunities generated through rural housing credit programs and (2) affordable rental opportunities for those not able to purchase a home

| Baseline 2009 | Target 2015 |
|---|---|
| 55,957 (homeownership); 469,162 (rental opportunities) | 222,000 (homeownership); 600,000 (rental opportunities) |

### Homeownership Opportunities

| 2006 | 2007 | 2008 |
|---|---|---|
| 42,172 | 43,532 | 66,574 |

1.2.3 Percentage of rural residents who are provided access to new or improved essential community facilities – (1) Health Facilities, (2) Safety Facilities, and (3) Educational Facilities

| Baseline 2009 | Target 2015 |
|---|---|
| Health: 5.4 percent (1.9 million people), Safety: 5.0 percent (2.6 million people), and Education: 3.5 percent (2.1 million people) | Health: 6.3 percent (3.8 million people), Safety: 5.8 percent (3.5 million people), and Education: 6.5 percent (3.9 million people) |

### Health

| 2006 | 2007 | 2008 |
|---|---|---|
| 3.8 percent | 7.2 percent | 4.8 percent |

### Safety

| 2006 | 2007 | 2008 |
|---|---|---|
| 3.8 percent | 6.2 percent | 5.7 percent |

1.2.4 Annual number of borrowers' subscribers receiving new or improved electric facilities

| Baseline 2009 | Target 2015 |
|---|---|
| 9,800,000[5] | 8,165,000 |

| 2006 | 2007 | 2008 |
|---|---|---|
| 8,200,000 | 5,800,000 | 8,100,000 |

---

[4] In the years 2008 and 2009, there were several loans to a borrower where the results were anomalies in reporting because those systems captured all the users in a large regional system.

[5] In 2009, there was $2.4 billion more available for program participants than is projected to be available in 2011. The 2015 target will be met with significantly less program-level funding.

## Strategies and Means

For developing rural community infrastructure, USDA employs strategies appropriate to the community's improvement. Perhaps the most important of these strategies is working in concert with other Federal, tribal, State, and local governments to develop partnerships to leverage resources in rural areas. The Department also encourages increased regional and community planning initiatives that build local planning capacity for taking advantage of economic and environmental opportunities. USDA provides economic analyses and promotes successful economic development models to regions to help facilitate strategic decision making to support livable communities. It may also assist in planning strategic investments that are community-driven in conjunction with a region's long-term vision. For instance, the Department works with State offices to improve priority performance goals by establishing place-based funding, decision-selection criteria, targeted areas, and customers.

USDA also coordinates outreach efforts supporting increased access to the Department's programs and services for women and minority farmers. The Department addresses the special needs of economically distressed regions by providing training to local staff and directing resources to projects in underserved and unserved communities. In addition, USDA programs provide educational and mentoring opportunities for youth that develop long term community leadership capacity.

To implement these strategies, USDA will:

- Facilitate the utility deployment of smart grid, renewable energy, energy efficiency, and carbon capture and storage for maintaining reliable electric systems in rural communities;
- Facilitate the deployment of water, waste, and environmental water systems;
- Facilitate the deployment of community facilities with critical health care, school, library, and safety investments to the neediest communities;
- Protect rural water supplies through watershed protection and restoration efforts;
- Improve access to green space and supporting livable communities through urban and community forestry programs;
- Collaboratively engage public lands communities in natural resource management;

- Engage tribal governments in tribal consultation concerning natural resource management;
- Work strategically with regions to promote sustainable planning and implementation; and
- Provide on-the-ground financial and technical support for regions.

## OBJECTIVE 1.3 – SUPPORT A SUSTAINABLE AND COMPETITIVE AGRICULTURAL SYSTEM

The economic vitality and quality of life in rural America and the U.S. economy at large depends on a financially healthy agricultural system. Agricultural is one of only a few sectors in the U.S. economy in which exports are creating a positive trade balance. U.S. agricultural producers are not simply farmers and ranchers. They are often small business owners trying to survive and support their families and rural communities in a challenging global, technologically advanced, and competitive business environment. USDA works to ensure that American farmers and ranchers are prosperous and competitive, have access to new and international markets, can manage their risks, and are supported in times of unusual economic distress or disaster. USDA also ensures agricultural and forestry land is used in an environmentally sustainable way.

It will take five coordinated tasks to meet Objective 1.3. These tasks are: ensuring a financially sustainable and competitive national agricultural system; facilitating access to international markets; supporting the development of new domestic markets; maintaining a strong financial safety net; and protecting the foundations of the agricultural system.

### Ensure a Financially Sustainable and Competitive Agricultural System

USDA management understands that, without profits and fair competition comparable to other U.S. business sectors, farm and ranch businesses cannot be sustained; the agriculture sector will not be able to produce the needed quantities of nutritious and safe foods, fuel, and fiber products for the rapidly growing U.S. and global populations; and rural communities cannot expect their youth to remain in farming. Thus, the Department plays a number of critical roles in increasing prosperity and sustainability in our Nation's agricultural system and rural communities. USDA effectively manages farm price support and commodity purchase programs to help balance supply and demand. The Department also uses oversight activities to protect producers from unfair competition and unfair business practices in the livestock, meat, and poultry markets. Working with the U.S. Department of Justice, the USDA helps prevent anti-competitive behaviors across the agribusiness sector.

USDA also assists producers in management and marketing by providing market trend analysis and business and marketing tools. The Department conducts research to improve performance and reduce the costs of agricultural inputs, including seed, fertilizer, and feed. USDA also looks for ways to improve crop and animal production practices, including environmentally sustainable ones. The Department further encourages the development of ecosystem markets that offer payments to producers for ecosystem services, such as carbon sequestration, water quality, wetlands, wildlife habitat, and species protection.

USDA works to ensure minority, women, beginning, and other socially disadvantaged farmers and ranchers have full knowledge and access to its programs. The Department also continues to work with producer and farmworker organizations and the U.S. Department of Labor to improve working conditions and income for farmworkers of every race and nationality.

### Facilitate Access to International Markets

For every $1 billion of agricultural exports, it is estimated that around 8,000 jobs are created and an additional $1.4 billion in economic activity is generated. One-third of all U.S. agricultural cash receipts come from export sales. In fiscal year (FY) 2010, U.S. agricultural exports are expected to reach $104.5 billion, up $7.9 billion from FY 2009. It is also the second-highest level on record behind $115.3 billion in FY 2008. USDA's Washington-based trade negotiators and technical experts work to promote agricultural trade. This work contributes directly to the prosperity of local and regional economies across rural America through increased sales and higher income.

The Department expands market opportunities in many ways. USDA connects agricultural exporters to customers, ensuring a level playing field for trade. It also provides timely information on agricultural markets in the United States and abroad, and facilitates trade through its network of overseas offices in 97 countries around the world. Cooperative efforts with other U.S. Government agencies and U.S. industry ensure that America's farmers and livestock producers have fair market access, a strong understanding of key market trends, and support in

overcoming market barriers. (See performance measures 1.3.3 and 1.3.4.)

USDA currently supports the farms-of-the-future concept, which describes strategies by which farmers and forest landowners can extend their production portfolios by providing and being compensated for measurable environmental benefits. USDA will support the development of domestic and foreign markets for ecosystem services, offering landowners an opportunity to take economic advantage of their natural resource assets and expand their portfolio to include private markets and receive payments for their environmental performance. The challenge will be to create new international market opportunities related to climate change. The Department must ensure U.S. negotiating positions are maintained in all international forums. Moreover, USDA offers capacity-building programs to its foreign partners that contribute to its goals, implements training and collaborative research programs, and monitors and shares climate change data.

### Support the Development of New Domestic Markets

One component of ensuring the financial sustainability of producers is to continue to identify and access new markets domestically. USDA provides support in developing opportunities through market trend analysis and business and marketing tools. This assistance includes overseeing national standards for the production and handling of agricultural products labeled as organic. (See performance measure 1.3.1.) Goods that are certified as organic frequently bring higher prices at market, resulting in increased returns for farmers. The Department also promotes access by producers to direct-to-consumer, local and regional, and other emerging opportunities. Finally, USDA supports the development of new markets, especially in food deserts and other underserved rural and urban communities.

### Ensure a Strong Farm Financial Safety Net

USDA helps maintain economic stability in the agricultural sector, enhancing the competitiveness and sustainability of farm economies. The Department strives to provide producers a backstop of prompt and equitable assistance, risk management tools, direct and counter-cyclical income payments, disaster assistance, and marketing assistance loans to farmers, ranchers, and eligible landowners. USDA also partners with commercial lenders to guarantee farm ownership and operating loans. It also makes direct loans to producers to purchase property or finance farm operating expenses.

The Department provides agricultural credit, among other services, to beginning farmers and ranchers, as well as those producers who traditionally have difficulty obtaining commercial credit. According to the 2002 Census of Agriculture, 273,000 minority, women, and beginning farmers were actively engaged[6] in the business of farming. In 2009, USDA helped finance 17.4 percent of this target group through direct and guaranteed loans. This number contrasts to USDA direct and guaranteed loans' 7-percent share of the overall U.S. agricultural credit market. The Department's financing for historically underserved farmers represented more than 40 percent of USDA's total loan portfolio in 2009. (See performance measure 1.3.2.)

Producers participating in the Federal Crop Insurance program determine the level of coverage they need to manage risks for their particular situations. This program allows them to mitigate production and revenue losses from yield or price fluctuations and receive timely claim payments. Producers can also use their individual crop insurance policy as collateral for commercial loans. USDA works collaboratively with other governmental and non-governmental organizations to provide risk management education, products, and services for producers. At this time, the crop insurance program has matured to a near-saturation level. The top 10 most commonly insured crops account for approximately 86 percent of total risk value, and approximately 80 percent of the planted acreage of these crops is insured. (See performance measure 1.3.5.)

### Improve the Foundations of the National Agricultural System

USDA efforts support the long-term viability of the U.S. agricultural system by providing Federal leadership in creating and disseminating knowledge spanning the biological, physical, and social sciences related to agricultural research, economic analysis, statistics, extension, and higher education.

The Department also conducts research to develop and transfer solutions to agricultural problems of high national priority. USDA provides access to and dissemination of information that ensures safe high-quality food and agricultural products. The Department's research efforts lead to more thorough assessments of the nutritional needs of Americans, new methods for sustaining a competitive agricultural economy, a better understanding of how to

enhance the natural resource base and the environment, and economic opportunities for rural citizens, communities, and society as a whole. USDA also supports the development of the next generation of agricultural scientists in the workforce. The Department works with higher-education institutions to develop education and training programs with strong science, technology, engineering, and math curriculums, and increase enrollment in secondary and two-year post-secondary education programs. For example, USDA will work to ensure support for agricultural research programs by doubling the number of students in agricultural science degree programs by 2020. (See performance measure 1.3.6.)

Additionally, the Sustainable Agricultural Research and Education (SARE) program funds projects and conducts outreach through competitive grants. SARE projects are designed to improve agricultural production systems by communicating how some farmers and ranchers cooperating on SARE grants have realized new ways to make money while protecting the environment and improving quality of life in their communities. The primary target audience for SARE includes America's two million small- and medium-sized farms and ranches.[7] SARE projects help farmers and ranchers improve their knowledge of sustainable agriculture production and marketing practices that ultimately leads to improved profitability, environmental stewardship, and quality of life. (See performance measures 1.3.7.)

### Performance Measures

1.3.1    Number of agricultural operations certified as organic

| Baseline 2009 | Target 2015 |
| --- | --- |
| 16,564 | 20,655 |

---

[6] $10,000 or more in gross sales.

[7] 2007 Census of Agriculture. Small- and medium-sized farms and ranches are less than 1,000 acres.

1.3.2 Percentage of Beginning Farmers, racial and ethnic minority farmers, and women farmers financed annually by USDA[8]

| Baseline 2009 | | Target 2015 | |
|---|---|---|---|
| 17.4 percent | | 18.5 percent | |

| 2006 | 2007 | 2008 |
|---|---|---|
| 15.5 percent | 15.9 percent | 16.2 percent |

1.3.3 Value of trade preserved annually through USDA staff intervention leading to resolution of barriers created by Sanitary/Phytosanitary (SPS) or Technical Barrier to Trade (TBT) actions

| Baseline 2006-2008 | | Target 2013-2015 |
|---|---|---|
| $4.1 billion[9] | | $4.5 billion |

1.3.4 Value of plant exports as a result of sanitary/phytosanitary agreements

| Baseline 2009 | | Target 2015 |
|---|---|---|
| $80 billion | | $104 billion |

1.3.5 Annual normalized value of risk protection provided to agricultural producers through the Federal Crop Insurance program

| Baseline 2009 | | Target 2015 |
|---|---|---|
| $53.9 billion[10] | | $55.9 billion |

| 2006 | 2007 | 2008 |
|---|---|---|
| $48.1 billion | $50.7 billion | $51.5 billion |

1.3.6 Number of students enrolled in undergraduate, graduate, and professional degree programs in agricultural sciences

| Baseline 2008 | | Target 2015 |
|---|---|---|
| 90,549 | | 113,186 |

1.3.7 Number of farmers and ranchers that gained an economic, environmental, or quality-of-life benefit from a change in practice learned by participating in a SARE project

| Baseline 2008 | | Target 2015 |
|---|---|---|
| 10,849 | | 14,300 |

| 2006 | | 2007 |
|---|---|---|
| 9,610 | | 10,240 |

## Strategies and Means

To increase the prosperity, competitiveness, and market opportunities for rural America, the USDA will:

Ensure a financially sustainable and competitive agricultural system

- Shape farm price support policies to increase farm and rural prosperity, and provide viable business opportunities for beginning farmers, racial and ethnic minority farmers and women producers;
- Provide agricultural producers and agribusinesses with information to make informed decisions in response to changing market conditions;
- Provide current, unbiased price and sales information to assist in the orderly marketing and distribution of farm commodities;
- Ensure quality price indexes are publicly available for all commodities that are included in the Farm Bill where a marketing loan is available or a USDA program payment can be made;
- Implement policy and regulations and perform industry analysis that keeps pace with the changing livestock, meat, and poultry industries;
- Enforce fair market practices and take action against anti-competition behavior, in partnership with the U.S. Department of Justice, to create a level playing field for producers;
- Ensure that USDA-approved and licensed warehouse programs maintain adequate storage facilities, adequate

---

[8] Many minority, women, and beginning farmers are not eligible for participation in USDA loan programs because they either qualify for commercial credit or do not need credit.

[9] Three-year rolling average.

[10] The 2009 baseline was modified from that included in the FY 2011 budget because a more current projection became available in the intervening time period.

frequency of warehouse examinations, and reduced product losses;

- Ensure that USDA foods are delivered timely, within contract specifications, and at competitive prices;
- Develop markets for ecosystem services and help farmers, ranchers, and forest landowners access those markets;
- Examine and accredit State and private certifying agents to ensure their compliance with national organic standards; and
- Focus on the specialized reporting of organic markets and continue to enhance the Market News program.

Facilitate access to international markets

- Provide agricultural producers with information to make informed decisions in response to changing market conditions both at home and abroad;
- Stimulate greater involvement in international markets by small- and medium-sized enterprises selling U.S. agricultural products;
- Promote increased sales by U.S. agricultural exports in rapidly growing foreign markets;
- Work to strengthen the global rules-based trading system upon which U.S. agriculture depends;
- Support international equivalency agreements with other countries;
- Facilitate the overseas marketing efforts of U.S. commodity organizations;
- Reduce technical barriers to trade and eliminate sanitary and phytosanitary barriers not based on sound science; and
- Ensure that international climate change agreements and measures are no more trade restrictive than necessary to meet the objectives of the agreement.

Support the development of new domestic markets

- Promote locally and regionally produced and processed food;
- Foster new opportunities for farmers and ranchers;
- Stimulate agriculturally and food-based community economic development;
- Expand access to affordable, fresh and local food;
- Cultivate healthy eating, and educated, empowered consumers; and

- Demonstrate the connection between food, agriculture, community, and the environment.

Ensure a strong farm financial safety net

- Improve the effectiveness of outreach efforts to minority producers, beginning farmers, and women by expanding efforts to partner with other Federal, State and local agencies, tribal governments, and non-governmental organizations that serve these targeted populations of agricultural producers;
- Enhance existing partnerships with land-grant universities and other educational organizations to identify and assist minority producers, beginning farmers, and women producers, and remove program barriers to participation;
- Collaborate with other governmental agencies, tribal governments, and non-governmental organizations to increase knowledge of risk management alternatives with a special emphasis on understanding the needs and barriers to utilization by minority, beginning, and women producers;
- Provide educational opportunities to beginning farmers and ranchers to help ensure they are knowledgeable of whole-farm planning, including farm financial management, conservation of natural resources, and increased agricultural productivity, and marketing opportunities, such as community-supported agriculture and farmers markets;
- Manage a fundamentally and actuarially sound crop insurance program that is growing its availability and product coverage, especially for livestock, pasture, rangeland, forage, and specialty crops;
- Inform farmers and ranchers of risk management tools and strategies, especially in underserved and emerging communities;
- Improve program compliance and crop insurance products through the increased use of technology, including geographical information systems, remote sensing, and data mining;
- Expand protection opportunities for producers through both price and revenue support, such as the Average Crop Revenue Election (ACRE) program;
- Mitigate the adverse results of natural disasters and provide relief to producers;
- Partner with other agencies to provide producers with information about Federal crop insurance, Noninsured

Crop Disaster Assistance Program (NAP), and other disaster assistance programs as they become available;

- Increase the use of Geographic Information Systems (GIS) to assess areas damaged by natural disasters and to speed delivery of disaster payments;

- Increase the capability of the enterprise-wide information technology infrastructure to support risk management solutions and farm program delivery;

- Expand outreach efforts for farm storage facility loans to eligible biomass, fruit, and vegetable producers;

- Facilitate the orderly marketing of major agricultural commodities through USDA Income Support programs by providing short-term financing or per unit revenue when market prices are low; and

- Continue to improve farm loan processing time and monitor repayment activity.

Improve the foundations of the national agricultural system

- Help higher-education institutions with undergraduate and graduate programs in agriculture develop strong science, technology, engineering, and math curriculums, and increase enrollment in secondary and two-year post secondary programs, especially from underrepresented groups;

- Advance production agriculture and sustainable agricultural systems through research that improves agriculturally important plants and animals, including plants and animals that are resilient to anticipated changes in climate;

- Advance knowledge to reduce greenhouse gas emissions in agriculture systems;

- Develop advanced biomass crops and methods for their sustainable production for use as biofuels and other forms of biopower;

- Develop these systems into viable and marketable products and ecosystem services;

- Develop new and improved practices to reduce producer costs;

- Use the cooperative extension system to transfer technology and best practices from the laboratory into active use; and

- Develop and maintain national standards governing the production and handling of agricultural products labeled as organic.

## External Risk Factors

Many external factors influence the outcome of this strategic goal. These might be changes in environmental conditions, including climate change, changing weather patterns, and ecosystem health. Other factors involve natural disasters, animal and plant pests and disease outbreaks, and intentional food contamination. Still others are primarily economic. Production-level factors include the volatility of farm commodity prices, workforce skills and competencies, and increasing input and operating costs for farms. Certain macroeconomic factors are also important, including rising unemployment, inflation, changes in the relative strength of the U.S. dollar to foreign currencies, and changes in the market demand for organic or bio-based products. Other influences include international concerns, such as trade policy and regulatory developments in other countries, and economic and cultural influences on student enrollment.

# Strategic Goal 2:
## Ensure Our National Forests and Private Working Lands[11] Are Conserved, Restored, and Made More Resilient to Climate Change, While Enhancing Our Water Resources

America's prosperity is inextricably linked to the health of our lands and natural resources. Forests, farms, ranches, and grasslands offer enormous environmental benefits as a source of clean air, clean and abundant water, and wildlife habitat. These lands generate economic value by supporting the vital agriculture and forestry sectors, attracting tourism and recreation visitors, sustaining green jobs, and producing ecosystem services, food, fiber, timber and non-timber products, and energy. They are also of immense social importance, enhancing rural quality of life, sustaining scenic and culturally important landscapes, and providing opportunities to engage in outdoor activity and reconnect with the land.

Federal, tribal, State and private lands face increasing threats from climate change, catastrophic wildfires, intense floods and drought, air and water pollution, aggressive diseases and pests, invasive species, and development pressures resulting in land and water conversion and reduced wildlife habitat. At the same time, there are immense opportunities to capture and increase the environmental, economic, and social benefits these lands provide.

USDA plays a pivotal role in protecting and restoring America's forests, farms, ranches, and grasslands while making them more resilient to threats and enhancing natural resources. The Department partners with private landowners to help protect the Nation's 1.3 billion acres of farm, ranch, and private forestlands. As public land stewards, USDA works to conserve and restore 193 million acres of National Forests and Grasslands in the National Forest System. The Department also partners with Federal, tribal, and State governments and non-governmental organizations to assist land and natural resource managers and connect people to the Nation's magnificent lands.

The Department is also a key player in the President's America's Great Outdoors initiative to build a 21st century conservation agenda, based on the experience and ideas of people from across the country. The goal of the initiative is to conserve our natural resources, both public and private, while reconnecting Americans to the outdoors.

USDA provides technical, financial, and planning assistance to its public and private partners. The Department's world-class data banks, research, and innovations give landowners and managers access to the latest science and technology to make informed decisions and implement conservation practices. USDA also connects forest and farm landowners with emergent markets for ecosystem services. This partnership allows landowners to reap the economic and environmental benefits of good stewardship.

The Department will use a collaborative, "all-lands" approach to bring public and private owners together across landscapes and ecosystems. The threats and opportunities facing the Nation's lands and natural resources do not stop at ownership boundaries. Over the next 5 years, USDA will help to restore and conserve the Nation's forests, farms, ranches, and grasslands (Objective 2.1); lead efforts to mitigate and adapt to climate change (Objective 2.2); protect and enhance America's water resources (Objective 2.3); and reduce the risk from catastrophic wildfire and restore fire to its appropriate place on the landscape (Objective 2.4).

### OBJECTIVE 2.1 – RESTORE AND CONSERVE THE NATION'S FORESTS, FARMS, RANCHES, AND GRASSLANDS

When the health and integrity of the Nation's lands deteriorate, so do the environmental, economic and social benefits they provide. This process carries enormous potential impacts on drinking water, greenhouse gas emissions, climate, wildlife, recreation, community health, and prosperity.

Restoring declining ecosystems and protecting healthy ones will ensure the Nation's lands are resilient to threats and impacts from a changing climate. It will also provide ecosystem benefits, food, fiber, and timber and non-timber products in a sustainable way. USDA will use the restoration of watershed and forest health as a core management objective of the National Forests and Grasslands. In many of the Nation's forests, restoration will include efforts to improve the health of fire-adapted or fire-impaired

---

[11] "Private working lands" include farms, ranches, grasslands, private forest lands, and retired cropland.

ecosystems, address the spread of insects and diseases that increase tree mortality, restore degraded wildlife habitat, improve or decommission roads, replace and improve culverts, and rehabilitate streams and wetlands. On agricultural and grazing lands, USDA will work with private landowners and managers to restore vegetative cover, rehabilitate streams and other water bodies, transition marginal or highly erosive lands to sustainable production levels, and apply conservation measures to enhance and maintain the quality of soil, water, and related natural resources. On all lands, the Department will actively work alongside its partners and landowners to conserve and invest in healthy ecosystems and watersheds to maintain their quality. USDA's collaborative "all-lands" approach will increase the scale and pace of restoration and conservation work on both public and private lands.

The Department is also working with farmers, ranchers, and forest landowners to maintain working lands and preserve open space. This work includes developing ecosystem markets and making strategic investments to purchase land or conservation easements. The latter is designed to slow, and in some cases reverse, the conversion and fragmentation of environmentally and economically significant forests, farms, and grasslands. Investments in urban and community forestry and other green spaces can also play a key role in increasing green space, improving ecological functions, and increasing ecosystem benefits.

### Performance Measures

2.1.1   Annual acres of public and private forest lands restored or enhanced

| Baseline 2009 | Target 2015 |
|---|---|
| 7.23 million acres per year | 8.15 million acres per year |

2.1.2   Acres of cropland with sustained productivity and improved ecological health

| Baseline 2003 | Target 2015 |
|---|---|
| 65 percent (260 million acres) | 70 percent (280 million acres) |

2.1.3   Percentage of non-Federal and USDA-managed grazing lands[12] with conservation or management applied to improve or sustain productivity and ecological health[13]

| Baseline 2009 | Target 2015 |
|---|---|
| 9.9 percent (66.25 million acres per year) | 10.5 percent (70.45 million acres per year) |

2.1.4   Total acres of agriculture and forest landscapes protected from conversion through conservation easements and fee simple purchases, to preserve natural resource quality, open space, and rural amenities

| Baseline 2009 | Target 2015 |
|---|---|
| 4.2 million acres | 6.5 million acres |

2.1.5   Number of communities with urban and community forestry programs resulting from Forest Service assistance

| Baseline 2008 | Target 2015 |
|---|---|
| 7,139 | 7,639 |

### Strategies and Means

To accomplish its restoration objectives and protect critical lands from conversion, USDA will:

- Develop collaborative strategies with landowners, State and local governments, other Federal agencies, tribes, and private sector organizations to address natural resource health and build community capacity to engage in natural resource work;

- Work with scientists, landowners, stakeholders, and partners to strategically identify and invest in the most environmentally and socially important landscapes;

- Ensure that National Forest System land management plans and projects are designed to restore degraded land and protect land that is healthy;

- Provide financial, technical, and planning assistance to communities and farmers, ranchers, and forest

[12] Total of 672 million acres.

[13] This includes the acres of grazing land on which conservation practices have been applied in the fiscal year, grazing land managed for conservation by the U.S. Forest Service, and acres enrolled in the Grasslands Reserve Program through the Farm Service Agency.

landowners to conserve, restore, and protect natural resources, and help them maintain and sustainably manage agricultural and forest lands;

- Strategically invest in conservation easements and land acquisitions to protect critical lands and leverage community investment in conservation practices;

- Identify vulnerable agricultural lands and work with producers to increase stewardship activities;

- Invest in urban and community forestry programs to restore urban landscapes and increase open space;

- Accelerate research and the development and implementation of tools and conservation practices that provide landscape benefits;

- Continue to invest in research to guide management practices; and

- Develop tools and materials to quantify the value of ecosystem services, monitor and assess conservation practice effectiveness, and connect and equip farmers, ranchers, and forest landowners with current market information and opportunities so they can earn revenue from the ecosystem benefits.

## OBJECTIVE 2.2 – LEAD EFFORTS TO MITIGATE AND ADAPT TO CLIMATE CHANGE

Climate change is one of the great challenges facing the United States and the world. The agricultural sector accounts for approximately 6 percent of total U.S. greenhouse gas emissions, while U.S. lands (primarily forests) absorb approximately 12 percent of total U.S. greenhouse gas emissions.[14] The Administration has pledged a 17-percent reduction in U.S. greenhouse gas emissions by 2020. Farmers, ranchers, and forest landowners can play a role in addressing climate change, and USDA is ready to help make that happen. The Department engages the participation of farmers, ranchers, and forest landowners in the Nation's efforts to reduce global warming.

The potential of our forest and agricultural lands to offset greenhouse gas emissions is significant. Plants absorb carbon dioxide from the atmosphere, store this carbon in sugars, starch, and cellulose, and release oxygen into the atmosphere. Healthy soils also hold and accumulate carbon.

---

[14] U.S. Environmental Protection Agency, *2009 U.S. Greenhouse Gas Inventory Report: Inventory of U.S. Greenhouse Gas Emissions and Sinks: 1990-2007* (April 2009).

Because of these natural processes, the sector represents a net sink of greenhouse gases. In fact, estimates show that forests in the United States could increase their absorptive capacity while being managed and restored for climate resiliency and future productivity.

USDA programs will help public and private land managers reduce greenhouse gas emissions and increase carbon sequestration on farms, ranches, and forestlands. These programs encourage a number of practices, including voluntary actions, offsets, and incentives. These practices include conservation tillage and precision nutrient management; improving energy and fertilizer efficiency; growing perennial grasses; planting trees on marginal farmlands, fire impacted landscapes, and built landscapes; minimizing deforestation; building methane digesters; facilitating gains in energy efficiency in agriculture and rural development; and using renewable sources of energy. USDA will also connect landowners to potential new markets for ecosystem services. For example, a potential greenhouse offset market could create economic value for landowners engaging in practices that reduce or sequester greenhouse gas emissions. In partnership with other agencies and departments, USDA could play a role in ensuring all offset markets have high standards of environmental integrity that result in real and measurable greenhouse gas reductions.

While reducing or sequestering emissions is one piece of the puzzle, USDA also provides leadership to help landowners and communities adapt to climate change impacts that are already appearing. These include changing water flow, availability, and quality; changing weather patterns and ambient temperatures; increased fire risk; increased threats from insects and disease; and changing habitat and climatic zones. The Department's land managers, conservation specialists, scientists, economists, and researchers play a leadership role in monitoring impacts and helping communities to adapt (e.g., measuring changes in water flow timing and intensity and then implementing relevant practices like placing stream buffers or upgrading culverts to absorb water overflow). USDA also forms partnerships to generate dialogue, coordinate investments, and facilitate regional planning.

The Department conducts research that will help inform decision-making about climate change policy, mitigation, and adaptation strategies. USDA research contributes to the development of climate change mitigation and adaptation tools and technologies. Its outreach and extension networks make these tools and technologies available to farmers, ranchers, and land managers.

## Performance Measures

2.2.1 Annual greenhouse gas emissions by the U.S. agricultural sector measured in CO2 equivalents (CO2 Eq.)[15]

| Baseline 2005 | Target 2015 |
| --- | --- |
| 482.6 million metric tons of CO2 Eq. | 441.6 million metric tons of CO2 Eq.[16] |

2.2.2 Annual amount of carbon sequestered on U.S. lands through voluntary actions, offsets, incentives, and actions on Federal lands

| Baseline 2005 | Target 2015 |
| --- | --- |
| 975.7 million metric tons of CO2 Eq. | 1,058.6 million metric tons of CO2 Eq. |

2.2.3 Percent of National Forests in compliance with a climate change adaptation and mitigation strategy

| Baseline 2009 | Target 2015 |
| --- | --- |
| 0 percent | 100 percent |

## Strategies and Means

USDA will work through its conservation and energy programs to reduce greenhouse gas emissions, sequester carbon, and lead adaptation efforts. An important part of the Department's mission is to measure, monitor, and validate the effects of conservation actions taken to reduce greenhouse gas emissions and sequester carbon to ensure their environmental benefits and economic viability. USDA will also collaborate with private landowners, Federal partners, tribes, States, local governments and external organizations to:

- Provide technical and financial assistance to farmers, ranchers, and forest landowners to implement conservation, nutrient management, and animal management practices that reduce emissions and sequester carbon;

- Develop and implement manure- and nutrient-management systems that reduce greenhouse gas emissions;

- Plant trees, grasses, or other appropriate vegetative cover and maintain existing vegetative cover on marginal farmland and land that has been impacted by fire;

- Increase green space in built environments and work to minimize conversions of working lands to non-agricultural and forest uses;

- Incorporate climate change mitigation and adaptation strategies into management practices and utilize scientific findings for all restoration projects, planning, and prescriptions;

- Include plans for protecting water quality and availability in climate change adaptation and mitigation strategies;

- Develop and assist entry to markets for greenhouse gas offsets to help landowners benefit from opportunities presented by climate change;

- Design, build, and operate USDA facilities in a sustainable manner, and invest in sustainable rural development and infrastructure;

- Develop renewable energy resources from sustainably managed farms, ranches, and forests;

- Invest in research and monitoring programs to increase the knowledge and understanding of how climate change impacts forest, crop, and range ecosystems;

- Contribute to periodic national assessments of the causes and consequences of climate change;

- Develop models, national observing and monitoring systems, decision support tools, and new technology and adaptation strategies for communities, agriculture producers, and natural resource managers;

- Develop science-based methods and technical guidelines for quantifying greenhouse gas sources and sinks in the forest and agriculture sectors;

- Track progress in meeting greenhouse gas reduction and carbon sequestration targets through rigorous national and regional greenhouse gas inventories of the forest and agriculture sectors;

- Build the capacity of the Department to measure the effects of greenhouse gas emissions and the effects of the actions to reduce them, to sequester carbon, and to create more opportunities for landowners;

[15] A carbon dioxide equivalent (CO2 Eq.) is a universal unit of measurement used to indicate the global warming potential of different greenhouse gases. One metric ton of CO2 Eq. has the same global warming potential as one metric ton of CO2.

[16] Interpolated based on a 17 percent reduction in total U.S. greenhouse gas emissions from 2005 levels by 2020.

- Expand plant and animal research to develop varieties and breeds that maximize carbon sequestration and can adapt to current and future climate conditions;

- Promote community involvement through listening sessions, symposiums, and workshops, and engage formal and informal education systems to share learning and improve climate preparedness;

- Engage in consultation, collaboration, and other techniques to improve tribal relations, involvement, and engagement relating to natural resource concerns;

- Encourage the adoption of reasonable, transparent, and science-based programs to adapt to, or mitigate the effects of, climate change on agriculture and forestry; and

- Advocate for U.S. agricultural climate change interests by engaging in discussion with foreign country officials and stakeholders.

## OBJECTIVE 2.3 – PROTECT AND ENHANCE AMERICA'S WATER RESOURCES

Protecting America's supply of clean and abundant water is among the most crucial environmental challenges of the 21st century. Water is essential for life. This precious resource is the foundation for healthy ecosystems, sustainable agricultural and forest production, livable communities, and viable industry.

Farmers, ranchers, and forest landowners play a pivotal role in protecting and enhancing water resources. Consider that 87 percent of America's surface supply of drinking water originates on our Nation's forests, farms, and range lands. The 193-million-acre National Forest System alone is the source of fresh water for more than 60 million people from coast to coast.

The restoration and protection of America's wetland ecosystems is important for protecting and improving water quality. This process also provides fish and wildlife habitats, stores floodwaters, and maintains surface water flow during dry periods. From the mid-1950s to the mid-1970s, the United States lost almost 500,000 acres of wetlands per year. This rate of loss was substantially reduced by 1997 and then eliminated by the first net gain in wetlands acreage in 2004. According to the Council on Environmental Quality, agricultural conservation and technical assistance accounted for 58 percent of the wetland acreage restored or created by Federal programs. Agricultural wetland restoration has

played a significant role in moving beyond a "no net loss" of wetlands. It continues to help increase the overall function and value of the Nation's wetlands.

USDA, in partnership with individuals, communities, and tribal, State, and local governments, implements programs that help ensure a clean and abundant supply of water for people and the environment. The Department provides technical and financial assistance for on-the-ground conservation and protects aquatic ecosystems and the headwaters of municipal water supplies in the Nation's National Forests and Grasslands. USDA also protects and restores wetland ecosystems, supports scientific discoveries, and engages in collaborative partnerships to maximize the benefits from public expenditures. Additionally, the Department provides assistance for community water infrastructure and water use efficiency investments. USDA leadership in supporting emerging water quality trading markets will provide new incentives to restore watersheds and wetlands, and manage agricultural lands for clean and abundant water supplies.

As water flows from forested headwater streams, through wetlands, over rural agricultural lands and urban watersheds to estuaries and oceans, USDA will strategically invest its resources to prioritize and accelerate the protection of water resources.

### Performance Measures

2.3.1 Acres of National Forest System watersheds at or near natural condition[17]

| Baseline 2009 | Target 2015 |
|---|---|
| 58 million acres (30 percent of National Forest System lands) | 62 million acres (32 percent of National Forest System lands) |

---

[17] Natural condition means that the watershed is unimpaired and stable. Some acres may require multiple treatments to vegetation (e.g. thinning, reforestation, hazardous fuels treatment, etc.), as well as structural investments (e.g. decommissioning roads, hazardous waste cleanup, structural improvements to roads, bridges, and culverts, etc.).

2.3.2 Acres of wetland ecosystems restored, enhanced, constructed, or protected on non-Federal lands in order to improve wetland functions and values, such as floodwater control, critical wildlife habitat, and carbon sequestration[18]

| Baseline 2009 | Target 2015 |
| --- | --- |
| 2.1 million acres per year | 2.3 million acres per year |

2.3.3 Value of flood prevention and water supply benefits delivered by rehabilitating or installing watershed structures and applying land treatment measures

| Baseline 2009 | Target 2015 |
| --- | --- |
| $2 billion in benefits delivered per year | $2.5 billion in benefits delivered per year |

2.3.4 Acres on which high impact targeted (HIT) practices are implemented on National Forest and private working lands in priority landscapes to accelerate the protection of clean, abundant water resources

| Baseline 2009 | Target 2015 |
| --- | --- |
| 0 acres | 9 million acres |

**Strategies and Means**

To protect and enhance water quality and availability and watershed health across landscapes, USDA will:

- Strategically focus investments in watershed improvement projects and conservation practices that will have the highest impact based on specific conservation needs within a given landscape;
- Accelerate the delivery of financial and technical assistance to farmers, ranchers, forest landowners, and producers to implement conservation measures and management strategies that benefit water quality and availability, improve water management, enhance water conservation, and protect and restore watershed health;

[18] This measure does not include the additional upland conservation work that would be undertaken to ensure proper wetland function. That upland work may be greater than the actual wetland area; some easements are 60 percent in uplands that are needed to ensure proper function of the targeted wetland system.

- Work with landowners to protect wetlands and other natural areas that help prevent flooding and storm surges from extreme weather events;
- Increase watershed-based collaborative partnerships with tribes, States, communities, landowners, and other stakeholders to build community-planning capacity and effectively guide development, protect and restore open space, and improve watershed management that supports clean and abundant water resources;
- Protect water resources on National Forest System lands by planning for watershed health and working to restore degraded watersheds, reduce erosion, reclaim and restore abandoned mine lands, reduce the threat of watershed damage from catastrophic wildfires, and reduce the impact of the road system on watershed health;
- Conduct a national assessment of water resource vulnerability to be used as a foundation for decisions on strategic investment of Department resources;
- Conduct research and provide financial and technical assistance to develop, promote, and deliver innovative technologies and science-based conservation and management practices to meet water quality and availability objectives; and
- Develop and implement a strategy for water quality trading through markets for ecosystem services to ensure that water benefits are valued in the marketplace and to mobilize private capital investments and individual donations for restoration activities.

OBJECTIVE 2.4 – REDUCE RISK FROM CATASTROPHIC WILDFIRE AND RESTORE FIRE TO ITS APPROPRIATE PLACE ON THE LANDSCAPE

Wildfires have a natural role on our landscape. Many forests, rangelands, and grasslands are actually dependant on fire for ecological health and sustainability. Despite this relationship, America's fire environment has changed. Some parts of the country are burning more frequently and with greater intensity than in the past. Impacts from climate change, drought, and the accumulation of flammable vegetation, combined with the increasing development in fire-prone areas, are causing more severe fires on the landscape and potentially increased damage to communities.

USDA works with Federal, tribal and State governments, and external partners to manage wildland fire, protect communities, reduce wildfire risk, promote community actions to address fire, and improve intergovernmental and interagency partnerships. Nearly 65,000 communities are at risk from wildfire across the United States. It is a high priority for the Department to work with these communities to reduce their risk from wildfire and help them take independent action to become "fire-adapted communities."

Preventing catastrophic wildfires is also important for ecological and watershed health. At the same time, restoring fire to natural areas that are fire-dependant ecosystems will help reduce fuel loads, lessen the risk of damaging fires, enhance wildlife habitats, and restore ecological and watershed function and resiliency. Thus, where appropriate, USDA will manage fire to safely restore it on the landscape.

The Department's ultimate goal is the safe, effective, and cost-efficient control of wildfires, while ensuring at-risk communities are fire-adapted. In addition, USDA strives to ensure fire is restored to its appropriate place on the landscape so that America's wildlands remain healthy and resilient.

### Performance Measures

2.4.1    Number of communities with reduced risk from catastrophic wildfire[19]

| Baseline 2009 | Target 2015 |
|---|---|
| 10,000 | 18,000 |

2.4.2    Cumulative number of acres in the National Forest System[20] that are in a desired condition[21] relative to fire regime[22]

| Baseline 2009 | Target 2015 |
|---|---|
| 58.5 million acres | 61.5 million acres |

---

[19] This is a joint performance measure with the Department of the Interior.

[20] Excluding Alaska, which does not have a consistent Fire Regime Condition Class data set.

[21] Defined as being within the natural (historical) range of variability of vegetation characteristics; fuel composition; fire frequency, severity, and pattern; and other associated disturbances.

[22] Fire regime is a generalized term for wildland fire's role within a vegetative community in the absence of modern human mechanical intervention (but including the influence of aboriginal burning) as characterized by fire frequency, predictability, seasonality, intensity, duration, and scale.

2.4.3    Percentage of acres treated in the Wildland-Urban Interface[23] that have been identified in community wildfire protection plans or equivalent plans

| Baseline 2009 | | Target 2015 | |
|---|---|---|---|
| 41 percent | | 55 percent | |
| **2006** | **2007** | | **2008** |
| 17 percent | 25 percent | | 36 percent |

### Strategies and Means

To accomplish the goal of reducing risks to communities and safely restoring fire to its appropriate role on the landscape, USDA, in partnership with the U.S. Department of Interior, and State and local agencies, will:

- Work with communities to develop, implement, and update Community Wildfire Protection Plans and improve local wildfire suppression capability and coordination;

- Develop, review, and update wildland fire management agreements to clarify jurisdictional interrelationships and define roles and responsibilities among Federal, tribal, State, and local fire protection entities;

- Improve fire decision support tools used on incidents to reduce risks to firefighters, increase efficiency, reduce costs, help restore and maintain sustainable landscapes, and protect communities;

- Collaborate with public and private forest and rangeland owners to develop and implement hazardous fuels reduction and ecosystem restoration projects to reduce the risk of catastrophic fire and make lands more resilient;

- Strategically and safely manage wildland fire and promote the appropriate use of prescribed fire to restore fire as a natural ecological process on the landscape, improve forest and habitat conditions, and reduce fuel loads and catastrophic fire risk;

- Implement USDA's American Reinvestment and Recovery Act projects and biomass utilization activities; and

---

[23] The Wildland-Urban Interface (WUI) is the area where undeveloped wildland vegetation meets or mixes with houses and human developments. The WUI is also where communities meet with wildland fuel, and therefore, the WUI is often the site of environment conflicts, such as the destruction of homes by wildfires.

- Conduct tribal consultation as discussions to further and enhance tribal collaboration.

## External Risk Factors

A number of outside factors affect USDA's ability to achieve Goal 2. These include: extreme weather, climate fluctuation, or environmental change beyond the natural range of variability that affects ecological productivity and resilience; increasing population, urban development and sprawl; increases in impervious surfaces and point and non-point source pollution beyond what the Department can influence through its programs; success of and level of participation in markets for ecosystem services; unpredictable economic fluctuations or commodity price changes that affect market conditions; budgetary, legal, and regulatory constraints; and international crises or homeland security issues that alter domestic program allocations or immediate public needs.

# Strategic Goal 3:

## Help America Promote Agricultural Production and Biotechnology Exports as America Works to Increase Food Security

A productive agricultural sector is critical to increasing global food security. For many crops, a substantial portion of domestic production is bound for overseas markets. USDA helps American farmers and ranchers use efficient, sustainable production, biotechnology, and other emergent technologies to enhance food security around the world and find export markets for their products.

Food insecurity is the lack of assurance that a person is always able to feed his or her family. Food security is measured by not only the availability of food but also the ability to purchase food. Food security means having a reliable source of food and sufficient resources to purchase it. A family is considered food secure when its members do not live in hunger or fear of starvation. Food security in foreign countries is affected by a number of factors, including the extent of the domestic food supply; the proportion of a nation's total volume of commodities used for such nonfood uses as feed or fuel; post-harvest losses due to waste and decay; the ability to finance food and agricultural imports; population income levels; and the proportion of income that must be devoted to food.

The Department is working to ensure U.S. agricultural resources contribute to enhanced global food security (Objective 3.1); enhance America's ability to develop and trade agricultural products derived from new technologies (Objective 3.2); and promote sustainable and productive agricultural systems that enable food-insecure nations to feed themselves (Objective 3.3).

### OBJECTIVE 3.1 – ENSURE U.S. AGRICULTURAL RESOURCES CONTRIBUTE TO ENHANCED GLOBAL FOOD SECURITY

Global food insecurity affects people worldwide, and the current global economic downturn only exacerbates the problem. Food assistance alone is not enough. Adequate food supplies must also be based on enhanced domestic, regional, and international trade, in-country increases in production, and the ability of the poor to earn sufficient incomes to purchase food. The largest contributing factors to insufficient in-country production are chronic under-investment in

agriculture, inefficient inputs and markets, and poor governance.

The United States has a strong interest in promoting sustainable agricultural systems in the developing world. Failing agricultural systems and food shortages fuel political instability in countries worldwide. This problem undermines global stability and threatens national security. Current examples of USDA's work to promote enhanced food security include its ongoing activities in Iraq and Afghanistan. In these countries, USDA employees are training local populations in state-of-the-art food preservation techniques, helping village populations develop local food supply chains from producer to consumer, and teaching local populations how to restore their watersheds. Programs like the McGovern-Dole International Food for Education and Child Nutrition Program expand traditional food assistance by targeting maternal, infant, and child nutrition programs in developing countries. McGovern-Dole also supports food security and a healthy young population, which in turn builds stable societies and increases national security. McGovern-Dole is the Department's largest international food-assistance program. It provides donations of U.S. agricultural products and financial and technical assistance for school-feeding and maternal and child nutrition projects in low-income, food-deficit countries committed to universal education. Due to hunger or malnutrition, an estimated 120 million school-age children around the world are not enrolled in school. McGovern-Dole projects help boost school enrollment and academic performance by providing school meals, teacher training, and related support. At the same time, nutrition programs are offered for pregnant and nursing women, infants, and pre-school youngsters. These programs sustain and improve the health and learning capacity of children before they enter school.

American agricultural resources and expertise play a significant role in increasing global food security by promoting technology- and science-based solutions and capacity-building activities in other countries.

## Performance Measure

3.1.1 Annual number of women and children assisted under McGovern-Dole International Food for Education Program

| Baseline 2009 | Target 2015 |
|---|---|
| 4.2 million | 5.0 million |

## Strategies and Means

To enhance global food security, USDA will:

- Assist in developing sustainable food systems in priority countries;

- Increase in-country capacity to develop sustainable agricultural systems;

- Promote the adoption of technology- and science-based solutions;

- Support U.S. global and national security policy through development and capacity-building activities to benefit other countries;

- Improve the Department's ability to respond to international crises involving food and agriculture;

- Supply decision- and policy-makers with timely market intelligence and analyses; and

- Promote sustainable agricultural production and improve U.S. national and food security through the Department's component of the Civilian Response Corps and the deployment of Department experts abroad to assist in developing sustainable food systems in priority countries.

## OBJECTIVE 3.2 – ENHANCE AMERICA'S ABILITY TO DEVELOP AND TRADE AGRICULTURAL PRODUCTS DERIVED FROM NEW TECHNOLOGIES

The United States is in a unique position to combat global hunger. U.S. farmers are among the most productive in the world. Additionally, domestic scientists are among the most advanced. The Nation produces bountiful supplies of staple foods like wheat, rice, corn, and soybeans. These products meet immediate food needs around the world through the development and promotion of new cutting-edge crop technologies and agricultural techniques.

USDA uses a science-based regulatory system. This system allows for the safe development and use of agricultural goods derived from new technologies that provide increased production options to agricultural producers. For example,

before a genetically engineered crop can be commercialized, the Department evaluates it thoroughly to ensure that it does not pose a plant-pest risk. This process ensures safe introduction and agricultural production options and enhances public and international confidence in these products. To further ensure confidence in these products, USDA will work with other Federal agencies to develop a comprehensive export strategy, including a plan for the continued export of products with expired patents.

Although the USDA and its Federal partners use a science-based regulatory system to develop and trade biotechnology products, trade with other nations is sometimes disallowed or interrupted without a valid, internationally recognized scientific justification. The Department actively works to advance science-based regulations and seeks to enforce existing global commitments governing trade in products derived through biotechnology. Additionally, USDA will work with its U.S. Government partners to help disseminate information related to the vast number of benefits of these technologies. The benefits of biotechnology products include addressing issues related to global food security, energy security, and climate change. Creating greater awareness of these benefits will lead to better acceptance of new technologies and increased trade opportunities for the Nation's farmers and ranchers.

## Performance Measure

3.2.1 Cumulative number of genetically engineered plant lines reviewed by USDA and found safe for use in the environment

| Baseline 2009 | | Target 2015 | |
|---|---|---|---|
| 80 | | 116 | |
| 2006 | 2007 | | 2008 |
| 70 | 73 | | 78 |

## Strategies and Means

To enhance the development and export of agricultural products derived from new technologies, USDA will:

- Regulate the importation, interstate movement, and field-testing of newly developed biotechnology-derived crops to ensure they do not pose a threat to plant health or the environment before they can be commercialized;

- Implement a coordinated strategy to facilitate the export of genetically engineered agricultural commodities;

- Coordinate responsibilities with the U.S. Environmental Protection Agency and the U.S. Food and Drug Administration as part of the Federal Coordinated Framework for the Regulation of Biotechnology;
- Engage international regulatory counterparts on science-based regulation;
- Work to advance internationally accepted science-based regulations with U.S. trading partners; and
- Ensure the enforcement of existing global commitments governing trade in agricultural biotechnology products.

## OBJECTIVE 3.3 – SUPPORT SUSTAINABLE AGRICULTURE PRODUCTION IN FOOD-INSECURE NATIONS

Agriculture is a powerful poverty-reduction tool. For every 1 percent of growth in agriculture in developing countries, poverty declines by as much as 2 percent. Because the majority of those who are hungry live in rural areas and depend on agriculture and natural resources for their livelihoods, investing in agriculture is the most efficient way to target those in need. Investment in the agricultural sector, including small-scale producers, contributes to overall economic growth. This investment increases efficiency in the marketing chain and reduces the share of income spent on food for poor populations. The Department promotes training and capacity-building programs for agricultural scientists, extension agents, and educators. The Department also promotes agricultural policy and regulatory programs that increase food safety and enhance a country's ability to take advantage of new trade and marketing opportunities.

USDA's work in Afghanistan provides an excellent example of the Department's efforts to support sustainable agricultural production in food-insecure nations. Agriculture is the main source of income for Afghanistan's economy. While only 12 percent of Afghanistan's total land area is fertile and less than 6 percent is currently cultivated, 80 percent of Afghanistan's population is involved in farming, herding, or both. The Department is helping Afghanistan revitalize its agricultural sector through a variety of activities designed to strengthen the Afghan government, rebuild agricultural markets, and improve management of natural resources.

A comprehensive approach to increase food supply in food-insecure nations must promote sustainable, market-led growth across the entire food production and market chain. This growth must occur from the laboratory to the farm to

the market to the table. It also must prevent and treat under-nutrition and increase the impact of humanitarian food assistance. Consumers must have access to an affordable food supply and assurance that the available food is nutritious and safe. USDA invests in capacity-building, technical assistance, and agricultural research to promote economic growth and increase agricultural production in developing countries. The Department also promotes agricultural policy and trade practices that enhance global food security, stabilize governments and societies, and enhance sustainable economic growth in developing countries.

### Performance Measure

3.3.1   Number of food-secure provinces in Afghanistan (34 provinces)[24]

| Baseline 2009 | Target 2015 |
|---|---|
| 10 | 16 |

### Strategies and Means

To promote sustainable agricultural systems in food-insecure nations, USDA will:

- Deploy agricultural experts from a wide range of USDA agencies to support the government of Afghanistan's agricultural programs throughout the country;
- Implement U.S. technical assistance, capacity-building, and agricultural support programs aligned with the U.S. Agricultural Assistance Strategy for Afghanistan, a whole-of-government strategy developed by the Office of the Special Representative for Afghanistan and Pakistan at the U.S. Department of State;
- Establish partnerships with international agricultural research centers, national agriculture research institutes in agricultural universities, relevant non-governmental organizations, and the commercial sector, with whom the Department has a common interest in improving crop and livestock production in food-insecure nations;
- Collaborate with partners to develop and release germplasm that enhances the nutrient content and

---

[24] Generally, throughout Afghanistan, food security conditions deteriorate during the winter, especially in chronically food-insecure zones. Food markets in these zones are seasonal and do not function year round. Thus, people live with food they have stocked prior to the arrival of winter.

productivity of crops and livestock in food-insecure nations and regions; and

- Invest in the research, development, and extension of new varieties and technologies to increase nutrient content value and productivity, and reduce pre- and post-harvest losses in major crops of interest to developing countries.

### External Risk Factors

America's ability to promote sustainable agricultural production practices and exports derived using biotechnology and other emergent technologies, while increasing food security at home and abroad, is affected by a number of external factors. These factors include: the failure of "change management" efforts to shift existing agricultural production technology practices toward more sustainable production methods; evolving scientific advances and industry practices; resistance both at home and abroad to foods produced through the use of biotechnology; the effects on worldwide food security from global climate change and the increasing variability of the weather; and political instability in food-insecure countries.

A plentiful is supply of safe and nutritious food is essential to the well-being of every family and the healthy development of every child in America. USDA supports and protects the Nation's agricultural system and the consumers it serves. The Department safeguards the quality and wholesomeness of meat, poultry, and egg products. USDA also provides nutrition assistance to children and low-income people who need it and addresses and prevents loss and damage from pests and disease outbreaks.

The Department works to improve the healthy eating habits of all Americans, especially children. Science has established strong links between diet, health, and productivity. Even small improvements in the average diet will yield large health and economic benefits. USDA's programs provide an infrastructure that enables the natural abundance of U.S. lands in combination with the ingenuity and hard work of the Nation's agricultural producers. This combination creates a food supply system unparalleled in its abundance, safety, and quality.

USDA helps put a healthy diet within reach of every American consumer by increasing access to nutritious food (Objective 4.1); promoting healthy diet and physical activity behaviors (Objective 4.2); protecting public health by ensuring food is safe (Objective 4.3); and protecting agricultural health by minimizing major diseases and pests to ensure access to safe, plentiful, and nutritious food (Objective 4.4).

### OBJECTIVE 4.1 – INCREASE ACCESS TO NUTRITIOUS FOOD

While most American households have access at all times to enough nutritious food for an active and healthy lifestyle, for too many in the U.S., hunger is a serious problem. The latest data show that 17 million American households, including more than 8 million households with children, had difficulty putting enough food on the table at some point during 2008. Even more alarming, in more than 500,000 households, with more than 1 million children, 1 or more children simply did not get enough to eat. They had to cut the size of their meals, skip meals, or go whole days without food at some time during the year.

The Administration has set a goal to end hunger among children by 2015. USDA's domestic nutrition assistance programs are critical to achieving this goal. These programs serve one in four Americans annually. In coordination with the Administration's anti-poverty strategies, these programs can potentially make hunger a thing of the past.

The Department is committed to providing benefits to every eligible person who wishes to participate in the major nutrition assistance programs, including the Supplemental Nutrition Assistance Program (SNAP), the Child Nutrition Programs, and the Special Supplemental Nutrition Program for Women, Infants, and Children (WIC). Too often people who need benefits do not participate because they may not know they are eligible, may not realize the size and value of benefits available to them, or may find applying too difficult or burdensome.

More than one-third of those eligible for SNAP are not participating. While the rate of program coverage has expanded in recent years, it remains below the record-high coverage levels of the 1990s. USDA targets a participation rate that exceeds those levels by 2015. The Department will use outreach to raise awareness of SNAP's benefits. USDA also will continue to work closely with State partners to streamline program operations and improve customer service.

USDA also intends to expand participation in the Child Nutrition Programs. The National School Lunch Program is available in most schools. Most public or nonprofit private schools of high school grade or under may participate in the school lunch program. Schools can receive cash subsidies and donated commodities from the Department for each meal they serve. In return, those schools must serve lunches that meet Federal requirements and offer free or reduced-price lunches to eligible children. Not all eligible children participate in the National School Lunch Program. Some bring healthful food from home, while others, especially in high schools, may be foregoing a nutritious lunch entirely.

Strategic Goal 4

The School Breakfast Program operates much like the National School Lunch Program, except the program may not be as reaching all the children who need it as effectively as its counterpart. Eating a healthy breakfast, at home or at school, is linked to better educational performance and classroom behavior, and fewer visits to the school nurse. Thus, USDA is expanding the program to ensure that it serves those who cannot otherwise get a nutritious breakfast. The Department is working to improve access to and the meal quality of these programs. These moves ensure that healthful and appealing food is available to every student to support growth and learning.

Promoting access to nutrition assistance goes hand-in-hand with managing these programs in a manner that ensures public confidence and maximizes the impact of Federal dollars. Strong management ensures that those most in need of nutrition assistance receive it. It also ensures that resources are not wasted by error or abuse. USDA uses all available opportunities, including new communication and eGovernment technologies, to serve customers, work with partners, and administer programs as effectively as possible.

## Performance Measures

4.1.1 Number of U.S. households with very low food security[25] among children, as measured annually with USDA's Food Security Supplement to the Current Population Survey

| Baseline 2008 | Target 2015 |
|---|---|
| 506,000 | 0[26] |

| 2006 | 2007 |
|---|---|
| 221,000 | 323,000 |

4.1.2 Annual percentage of eligible people participating in SNAP

| Baseline 2007 | | Target 2015 |
|---|---|---|
| 65.8 percent | | 75.0 percent |

| 2004 | 2005 | 2006 |
|---|---|---|
| 61.1 percent | 64.7 percent | 67.3 percent |

4.1.3 Annual percentage of eligible people participating in the National School Lunch Program

| Baseline 2009 | | Target 2015 |
|---|---|---|
| 56.2 percent (31.2 million participants) | | 60.0 percent (34.5 million participants) |

| 2006 | 2007 | 2008 |
|---|---|---|
| 54.6 percent | 54.9 percent | 55.5 percent |

4.1.4 Annual percentage of eligible people participating in the School Breakfast Program

| Baseline 2009 | | Target 2015 |
|---|---|---|
| 19.8 percent (11.0 million participants) | | 25.0 percent (14.4 million participants) |

| 2006 | 2007 | 2008 |
|---|---|---|
| 9.8 million | 10.1 million | 10.6 million |

## Strategies and Means

To increase access to nutritious food, USDA will:

- Increase participation rates in nutrition assistance programs through aggressive and creative outreach, customer service improvements, earned media activities, demonstration projects, and research and analysis to identify reasons for and potential solutions to participation gaps;
- Invest in research to develop innovative program improvements, measure progress, and build a solid foundation to inform future policy, outreach, and partnership initiatives;
- Work with tribes and States to streamline applications, use integrated technology and business process improvements, and make operational changes to programs and policies to facilitate easier access to nutrition assistance programs, especially for children;
- Support tribal and State efforts to increase the use of electronic benefit transfer technology in WIC;

[25] Very low food security is the most severe form of limited food access. In households with very low food security among children, one or more children do not get enough to eat—they had to cut the size of their meals, skip meals, or even go whole days without food at some time during the year.

[26] In practice, a result of 80,000 or fewer households with very low food security among children is indistinguishable from zero, given the precision of the survey sample.

- Use Child Nutrition Reauthorization to improve program access and expand eligibility for children in need;

- Support tribal and State efforts to improve SNAP benefit accuracy through oversight, training, technical assistance, and "promising practices" information sharing;

- Expand support for research on nutrient density and diet through both intramural and extramural research;

- Develop and equip grocery stores and other small businesses and retailers to sell healthy food in communities that currently lack these options, thereby improving food options, creating jobs, helping revitalize distressed communities, and opening up new markets for farmers to sell their products, which can provide an economic boost to rural America; and

- Promote public-private partnerships to encourage grocery store and other healthy food outlet development in underserved areas, helping tackle the obesity epidemic, creating jobs, and revitalizing low-income neighborhoods.

## OBJECTIVE 4.2 – PROMOTE HEALTHY DIET AND PHYSICAL ACTIVITY BEHAVIORS

Science has established a clear linkage between proper diet, adequate physical activity, and good health. Unfortunately, for too many in the United States, poor dietary habits and a lack of physical activity contribute to devastating health conditions, including overweight and obesity, coronary heart disease, hypertension, and heightened risk of stroke, diabetes, and some cancers.[27]

Perhaps the most troubling of these conditions is the epidemic rate of childhood obesity. Over the last 30 years, the childhood obesity rate has more than tripled.[28] This has serious implications for the Nation's future. Children who start out life obese have greater struggles with their weight in later years. In fact, 80 percent of teenagers who are obese remain obese as adults and at risk for the chronic diseases described above. Beyond the physical suffering such diseases cause, obesity-related

---

[27] Centers for Disease Control and Prevention Web site, "Overweight and Obesity – Health Consequences" (http://www.cdc.gov/obesity/causes/health.html)

[28] U.S. Department of Health and Human Services. *The Surgeon General's Vision for a Healthy and Fit Nation.* Rockville, MD: U.S. Department of Health and Human Services, Office of the Surgeon General, January 2010, p. 2.

health complications can increase medical costs to America's health care system and decrease overall productivity.

The Administration has set a goal to solve the problem of childhood obesity within a generation so that children born today will reach adulthood at a healthy weight. The First Lady has launched the *Let's Move!* campaign to lead this effort. USDA has critical roles to play in its success. While the Department will work collaboratively with other Federal agencies on a wide array of strategies to reduce childhood obesity and improve diets, USDA will focus on areas where the Department's unique strengths and capabilities can have the greatest impact.

USDA is fighting to reverse the rapid increase in childhood obesity by improving school meals and the school nutrition environment. On school days, children who participate in both the breakfast and lunch programs consume as many as half of their calories at school. The Department must ensure that all foods served in school contribute to good health. USDA will update school meals standards based on recommendations by the Institute of Medicine. The Department also will ensure that schools have the skills to serve top-quality, healthy, and appealing meals. USDA will create national baseline standards for all foods sold in elementary, middle, and high schools to ensure they contribute effectively to a healthy diet.

As part of this effort, USDA is encouraging and supporting WIC mothers to breastfeed their infants by strengthening breastfeeding policy and program activities. Breastfeeding is linked to a lower risk of numerous health problems for both mother and infant. It also may have a protective effect against pediatric overweight and obesity. Breast milk provides the best source of infant nutrition and helps infants get a healthy start in life. USDA will recognize and reward State achievements in promoting breastfeeding through performance awards, expand the availability of peer-counseling in WIC clinics, and continue its core promotion and support activities.

Furthermore, USDA is working to double the number of Americans who eat five or more servings of fruits and vegetables daily. Too many Americans simply do not eat enough of these critical foods to support good nutrition. The Department boasts evidence-based nutrition guidance, including the *Dietary Guidelines for Americans* (DGA), which was developed in partnership with the U.S. Department of Health and Human Services, and the

*MyPyramid* food guidance system. These tools are designed to give every consumer the knowledge and motivation they need to choose a diet that supports a healthy future. USDA will use these tools to promote fruit and vegetable consumption and other healthy eating behaviors through public-private partnerships and cutting-edge technologies.

The Department is also working to improve access to healthful, locally produced food, especially for nutrition-assistance clients. Americans' food choices are strongly influenced by the choices most easily available to them. Limited access to nutritious food and relatively easier access to less nutritious food may be linked to poor diets. USDA will improve access to healthy, locally produced foods, such as fruits and vegetables. The Department also will increase opportunities for farmers and food entrepreneurs to retail their produce in low-access areas by encouraging farmers markets in those areas to become authorized SNAP vendors, equipping markets to accept SNAP electronic benefit transfer (EBT) benefits, and promoting greater use of farmers markets to SNAP participants. Through its farm-to-school efforts, USDA also connects schools with regional and local farms to serve healthy meals using locally produced foods. The Department's People's Garden Initiative also creates places where young people can learn to grow their own nutritious food.

To improve Americans' physical activity behaviors, USDA is promoting the use of the Nation's public lands to increase healthful physical activity. Most Americans need to move more to promote their health and well-being and improve the "energy balance" between the calories they consume and those they expend. National forests and grasslands are America's backyard, offering chances to increase physical activity. Nearly 100 million National Forest System visitors participate in outdoor-based physical activities each year. USDA will promote the use of these lands and facilities to nurture the body and mind in natural settings near local communities. The Department also will increase access to green space through such programs as the Forest Service's Urban and Community Forestry program.

These efforts will help America's children develop lifelong healthy habits, improve nutrition and well-being for all Americans, and build a healthier and more prosperous future for the Nation.

**Performance Measures**

4.2.1 Percentage of children and adolescents who are obese[29]

| Baseline 2008 | Target 2015 |
|---|---|
| 16.9 percent | 15.5 percent |

4.2.2 Percentage of WIC mothers that breastfeed

| Baseline 2008 | Target 2015 |
|---|---|
| 59 percent | 65 percent |

4.2.3 Percentage of adults who consume five or more servings of fruits and vegetables daily

| Baseline 2007 | Target 2015 |
|---|---|
| 24.4 percent | 50.0 percent |

4.2.4 SNAP benefits redeemed at farmers markets annually

| Baseline 2008 | Target 2015 |
|---|---|
| 753 markets authorized; $2.7 million in benefits redeemed | 2,000 markets authorized; $7.2 million in benefits redeemed |

**Strategies and Means**

To promote healthy diet and physical activity, USDA will:

- Update and implement improved school meals nutrition standards based on Institute of Medicine recommendations to conform to DGA; complete proposed and final rulemaking within 2 years; work with Congress to increase State program resources conditioned on improved meal quality; and provide technical assistance and support to implementing schools;

- Provide parents and students better information about school nutrition and meal quality;

- Secure statutory authority and set standards that increase consistency with DGA for all foods sold in schools;

- Improve the school nutrition environment by expanding the *HealthierUS School Challenge* and creatively promoting DGA- and *MyPyramid*-based messages;

[29] Figures represent children ages 2-19 with Body Mass Index (calculated from height and weight) who measure at or above the 95th percentile for their age and sex.

- Improve the quality of other foods offered and sold at schools under current authority by doubling the number of schools meeting the *HealthierUS Schools Challenge* during the 2010-2011 school year, and adding 1,000 more annually for each of the 2 years after that;

- Work with school business administrators to share best practices and revenue strategies that do not entail sales of food that do not meet standards;

- Encourage food manufacturers and marketers to develop new products and reformulations that meet national standards and appeal to children;

- Promote the increased consumption of whole grains, fruits and vegetables, and low-fat and fat-free milk through the application of the principles of behavioral economics in school cafeterias;

- Promote the increased consumption of fruits and vegetables among the general population;

- Support and encourage breastfeeding, the medically preferred feeding practice for most infants, through WIC and other venues;

- Expand intramural and extramural behavioral research into obesity, especially among children;

- Evaluate nutrition-promotion interventions to implement and sustain evidence-based strategies in communities across the Nation;

- Increase access to locally grown fruits and vegetables and other nutritious food by expanding the use of SNAP electronic benefit transfer in farmers markets, among other strategies;

- Use national public television to educate viewers on DGA;

- Use the USDA People's Garden Initiative to teach young people how to grow their own nutritious food;

- Collaborate with the U.S. Departments of Health and Human Services and the Treasury to increase the availability of affordable, healthy foods in underserved urban and rural communities;

- Encourage outdoor exercise and recreation on public lands through partnerships, outreach campaigns, and recreational programs; and

- Increase the number of children and adults who have access to green space for physical and mental health.

## OBJECTIVE 4.3 – PROTECT PUBLIC HEALTH BY ENSURING FOOD IS SAFE

USDA is committed to ensuring Americans have access to safe, nutritious, and balanced meals. Today, as many as one in four Americans experience a foodborne illness annually.[30] The Department is working to significantly reduce that number. USDA takes a farm-to-table approach to reducing and preventing foodborne illness. The Department invests in its workforce and data infrastructure to prevent harm to consumers, minimizing the prevalence of food contaminants and quickly identifying and averting outbreaks. Effective food safety inspections and enforcement depend upon timely quality data and analysis.

USDA protects public health by ensuring that meat, poultry, and processed-egg products are safe, secure, wholesome, and correctly labeled and packaged. The Department annually inspects 63.1 billion pounds of processed products. From these USDA-regulated products in fiscal year 2009, there were an estimated 562,697 illnesses reported from *Salmonella, Listeria monocytogenes*, and *E. coli O157:H7*. This number is the equivalent of 1 illness for every 298,796 servings of meat and poultry products consumed.[31]

The Department is working to better verify that effective food safety systems are correctly operating in USDA-regulated slaughter and processing establishments. USDA has invested in upgrading and training inspection personnel, developing an automated system to alert inspectors about potential food safety problems, and giving inspectors greater and more timely access to establishment performance data.

USDA also measures industry adoption of functional food defense plans. Food defense plans are written procedures that food processing establishments should follow to protect the food supply from intentional contamination with chemicals, biological agents, or other harmful substances. These plans help the agricultural industry protect public health and reduce negative economic impacts on the food infrastructure.

---

[30] Estimate of prevalence illnesses based on 76 million annual domestically acquired foodborne illnesses, Mead et al. (1999). Mead PS, Slutsker L, Dietz V, McCaig LF, Bresee JS, Shapiro C, Griffin PM, and Tauxe RV. "Food-related illness and death in the United States." *Emer Infect Dis*: 1999, 5(5):607-2.

[31] Illnesses per serving calculated using FY2008 ERS data on U.S. per capita loss-adjusted food availability for all red meat and poultry products and the FSIS All-Illness Measure from FY 2008 as ERS had not published data on U.S. meat consumption for FY 2009 when this document was written.

Additionally, at the retail level, USDA's outreach efforts help alert consumers to food safety recalls. The Department is hiring more epidemiologists to coordinate with State officials to develop "trace back" tools and improve outbreak identification and response.

Finally, as imported products and on-farm practices can dramatically impact food safety, USDA is developing inspection strategies based on public health decision criteria to ensure import safety and provide guidance to promote good agricultural practices on the farm.

The Department will use all available data along the farm-to-table continuum to target its resources effectively to prevent outbreaks and restrict them when they occur; inform the development of policies and risk management decisions; and evaluate the effectiveness of its initiatives. In addition, with the launch of the Public Health Information System, USDA will be analyzing its data in real-time to identify potential food safety risks in the food supply and rapidly respond.

## Performance Measures

4.3.1 Percentage of broiler establishments found to be in Category 1 as determined by *Salmonella* verification testing[32]

| Baseline 2009 | Target 2015 |
| --- | --- |
| 82 percent | 97 percent |

| 2007 | 2008 |
| --- | --- |
| 73.5 percent | 83 percent |

4.3.2 Total number of *Salmonella, Listeria monocytogenes*, and *E. coli* O157:H7 illnesses from products regulated by USDA's Food Safety Inspection Service[33]

| Baseline 2009 | Target 2015 |
| --- | --- |
| 562,697 | 505,024 |

4.3.3 Average percentage of all establishments with a functional food defense plan[34]

| Baseline 2009 | Target 2015 |
| --- | --- |
| 62 percent[35] | 90 percent |

## Strategies and Means

To protect public health and ensure food safety, USDA will:

- Address the education, outreach, and food safety challenges identified by the President's Food Safety Working Group;

- Monitor and evaluate data from field operations to identify new and emerging pathogens and sources of contamination, and release updated notices and directives to field personnel and regulated establishments to ensure the most updated and accurate procedures are utilized as new issues arise;

- Verify through the Department's inspection activities and through routine and "for-cause" food safety assessments (FSAs – comprehensive assessments of food safety plans) that establishments are following their

---

[32] Establishments are placed in Category 1 if they demonstrate consistent process control in USDA verification testing. Category 1 is the highest measure attainable by establishments. USDA plans to tighten its *Salmonella* performance measure for broilers, which will require a downward adjustment to a lower percentage attained for the FY 2015 performance goal for *Salmonella* broiler carcasses in the first quarter of FY 2011.

[33] Both the 2009 baseline and the FY2015 target were modified from what was included in the Agency's FY2011 budget as updated attribution estimates became available in the intervening time period and the Agency shifted to using new CDC FoodNet data rather than older estimates of foodborne illness burden. USDA will update its attribution estimates on an annual basis, and, as estimates are not likely to significantly change from year to year, performance objectives will be re-evaluated on a three-year cycle.

[34] Food defense plans are written procedures that food processing establishments should follow to protect the food supply from intentional contamination with chemicals, biological agents or other harmful substances. To be functional, an establishment must develop, write, implement, test, assess, and maintain the food defense plan.

[35] The 2009 baseline was modified from that included in the FY 2011 budget because updated survey data on industry food defense adoption rates became available in the intervening time period.

Hazard Analysis and Critical Control Points (HACCP) plans;

- Target additional FSAs in poor-performing establishments to analyze an establishment's control of *Salmonella* and the design and implementation of an establishment's food safety system;

- Implement the Public Health Information System (PHIS) to maximize the performance of food safety verification and sampling procedures performed by 3,000 USDA inspection program personnel and 800 inspection program personnel in 27 state programs, computing results into real-time predictive analytics to target poor performing plants and initiate intensified inspection and enforcement;

- Update sampling strategies as new technologies and methodologies are developed, and seek to ensure sampling plans include all high-risk products;

- Ensure that all new policies and procedures are data-driven and consider the most up-to-date science;

- Continue a proactive dialogue with industry and all food safety partners to seek their participation on how to best address hazards posed by food the Department regulates;

- Reach out to and forge stronger partnerships with other Federal and USDA agencies that deal with producers and consumers;

- Modernize the Department's workforce through the addition of field investigators to maintain a lead investigative role in foodborne illness investigations and interaction with public health partners to identify adulterated products, support recalls, and initiate enforcement actions;

- Train Enforcement, Investigation, and Analysis Officers (EIAOs), to ensure that all employees have access to the newest and most effective data and methodologies;

- Support new and innovative initiatives, such as the *Salmonella* Initiative Program (SIP), to reduce pathogens in the foods the Department regulates by driving improvements in on-going control for *Salmonella* in broiler and turkey slaughter operations;

- Initiate baseline studies on chicken parts, market hogs, and pre-pasteurized egg products, using collected baseline data to analyze trends and relationships between pathogen levels and drive new performance standards;

- Participate in research with Departmental, State, tribal, local, community, and academic partners, such as product-focused and epidemiology studies, to ensure the best science is being used to support Departmental policies and programs;

- Educate consumers via *Ask Karen* and the Meat and Poultry Hotline on safe-cooking practices to reduce and prevent the improper cooking and handling of ground beef and other USDA-regulated products; and

- Provide training and educational materials for all regulated processing plants on food safety and the design of food-processing facilities.

## OBJECTIVE 4.4 – PROTECT AGRICULTURAL HEALTH BY MINIMIZING MAJOR DISEASES AND PESTS TO ENSURE ACCESS TO SAFE, PLENTIFUL, AND NUTRITIOUS FOOD

USDA helps keep safe, nutritious food accessible and affordable by preventing the entry and establishment of agricultural pests and diseases and minimizing production losses. Safeguarding animal and plant resources against the introduction of foreign agricultural pests and diseases provides access to a diverse supply of fruits, vegetables, meat, and poultry. The Department detects and quickly responds to new invasive species and emerging agricultural and public health situations. The Department's strategy involves (1) identifying threats overseas and working to prevent them from coming to the United States, (2) providing training and expertise to identify threats at ports of entry, and (3) working to eradicate pests and diseases or manage them to limit the damage done if they are already in the United States. Where possible, USDA eradicates or manages existing agricultural pests and diseases and wildlife damage. The Department also develops and applies more effective scientific methods to prevent, detect, eradicate, or manage pests and diseases. These efforts contribute to the overall agricultural health of the Nation and the world.

### Performance Measure

4.4.1  Value of damage prevented and mitigated annually as a result of selected plant and animal health monitoring and surveillance efforts

| Baseline 2009 | Target 2015 |
|---|---|
| $1.05 billion | $1.67 billion |

| 2007 | 2008 |
|---|---|
| $1.37 billion | $1.38 billion |

## Strategies and Means

To help minimize the damaging effects of pests and diseases on the Nation's food supply, USDA will:

- Reduce or eliminate "quarantinable" pests and animal diseases that can cause significant economic damage;

- Facilitate trade (resolve scientific trade barriers, conduct preclearance programs, etc.) to provide a wide variety of nutritious, fresh fruits and vegetables year-round;

- Work with foreign governments and agricultural health expert counterparts to control plant and animal pests and diseases;

- Collaborate with State governments, local officials, and others on the early detection and emergency response efforts to find pests and diseases, should they reach the United States, and determine an appropriate course of action to prevent their spread;[36]

- Develop control and eradication programs for pests and diseases that have become established, considering the science and technological tools available, and with the assistance of State governments and industry participants; and

- Develop methods to address pests and diseases of concern, often working with universities, and provide diagnostic support to enhance pest and disease programs.

## External Risk Factors

The USDA's ability to ensure that all Americans have access to safe, nutritious, and plentiful food supplies is impacted by several external factors. These factors include: the effectiveness of State and local organizations that deliver benefits for nutrition-assistance programs; the collaborative efforts of other Federal agencies that deliver or support health, human services, and education benefits; problems with food handling or preparation that lead to outbreaks of foodborne illness; increases in the volume and types of food products available on the market; food terrorism and intentional contamination and infestation of the food supply; changing human consumption trends; gaps in food safety record-keeping by outside parties; and increased risks of pest and disease introductions through globalization and more open trade practices.

---

[36] For status on controlling current infestations visit http://www.aphis.usda.gov/newsroom/hot_issues/index.shtml.

# Management Initiatives

## OVERVIEW OF MANAGEMENT INITIATIVES

USDA is working to transform itself into a model organization. By strengthening management operations and engaging employees, the Department will improve customer service, increase employment satisfaction, and develop and implement strategies to enhance leadership, performance, diversity, and inclusion. The transformation will result in process improvements and increased performance.

USDA expects to:

- Transform itself into a model Federal department for effective program delivery by enhancing leadership, encouraging employee inclusion, and focusing on improving customer and employee satisfaction;
- Provide civil rights leadership to its employees, applicants, and customers by reducing the inventory of program civil rights complaints, analyzing field operations for systemic improvements, and increasing the use of early resolution processes (a form of alternative dispute resolution) for civil rights and equal employment opportunity complaints;
- Coordinate outreach efforts to increase access to its programs and services among women and minority farmers;
- Use resources more effectively by incorporating new strategies and policies into its management practices that increase performance, encourage efficiency, and align activities to the Department's strategic goals;
- Implement modern information technology systems and policies in a cost effective manner that improve program delivery and internal and external communications capabilities to better serve USDA constituents;
- Maximize its "green" operations by increasing recycling and the use of bio-preferred products and alternative energy, and decreasing water and energy usage at its facilities;
- Improve Departmental emergency preparedness and security measures to protect its employees and the public to ensure the continued delivery of its products and services; and

- Enhance human resources policies and practices to develop a workforce more representative of the national population and that has the necessary skills to ensure the continued and improved delivery of services.

## Initiative I: Engage USDA Employees to Transform USDA into a Model Agency

Engaging employees to transform USDA into a high-performing, inclusive department that benefits from leadership at all levels will require a pointed and comprehensive effort. USDA will continue to generate opportunities to listen to employees' concerns and ideas. The Department also will design and implement beneficial systemic changes to processes that affect employee satisfaction and human resources.

USDA's plans include:

- Developing and implementing comprehensive strategies to improve leadership;
- Effectively managing employee development, talent management, employee progression, and customer and community outreach across the Department; and
- Measuring and increasing the satisfaction of its customers and employees.

## Initiative II: Provide Civil Rights Services to Agriculture Employees and Customers

USDA leadership has established civil rights as one of its top priorities. To be successful, all employees must be committed to making the Department a model in the Federal Government for respecting the civil rights of its employees and constituents. USDA will change the direction of equal employment opportunity, civil rights, and program delivery through a comprehensive approach. This approach will ensure fair and equitable treatment of all employees and applicants. It also will improve program delivery to every person entitled to services. This effort will assist the Department to address past errors, learn from its mistakes, and move forward to a new era of equitable service and access for all.

USDA's plans include:

- Increasing early resolution usage in program civil rights and equal employment opportunity complaints;
- Reducing the inventory of program civil rights complaints; and
- Analyzing field operations for systemic improvements.

## Initiative III: Coordinate Outreach and Improve Consultation and Collaboration Efforts to Increase Access to USDA Programs and Services

USDA will coordinate and measure the performance of strategic outreach efforts to ensure that all Americans have equal and fair access to key Department programs and services. By promoting USDA values and priorities in such efforts as the People's Garden, the Department looks to enhance the public's knowledge of sustainable growing practices and the importance of the American farm community. USDA will develop or expand enterprise-wide, results-driven initiatives and coordinated efforts. This collective work will increase the viability and profitability of small farm operators and beginning and socially disadvantaged farmers and ranchers. The Department will coordinate various USDA services for such priority populations as farm workers. USDA also will improve its compliance with Executive Directives by requiring consultation and improved collaboration with tribal governments. Significant improvements will be made in USDA's consultation processes, policies, reporting, and outcomes.

USDA's plans include:

- Measuring and increasing participation in key programs among small and beginning farm operators and socially disadvantaged farmers and ranchers;
- Coordinating and enhancing programs and services for farm workers to ensure the stability of the agricultural labor force;
- Simplifying procurement and program application processes to ensure fairness and equity of opportunity;
- Expanding the People's Garden concept to exhibit Department values across the Nation;
- Building and leveraging partnerships between itself and non-profit and faith-based organizations to better serve individuals, families, and communities;
- Launching the Office of Tribal Relations, instituting Departmentwide policies and reporting procedures, and

launching regionally based technical assistance and dialogue opportunities for tribal governments, tribal communities, and individuals served by tribal governments.

- Enhancing tribal relations through improved consultation in compliance with President Obama's presidential proclamation, related directives, and Executive Orders, and the use of targeted means to reach tribal governments, communities, and individual tribal members;
- Increasing USDA commodity procurement contracting opportunities for small-farmer-owned cooperatives through contracting outreach and education, and the aggressive use of contracting vehicles, such as 'set-asides' and preferences; and
- Improving the health and wellness of Federal employees by serving local and nutritious food at USDA cafeterias.

## Initiative IV: Leverage USDA Departmental Management to Increase Performance, Efficiency, and Alignment

In a world of increasingly tight budgets and expanding responsibilities, it is necessary for USDA to ensure the effective and efficient use of its resources. The Department must maintain effective financial controls so that program dollars achieve the outcomes for which they were intended. USDA, led by Departmental Management, will implement strategies to strengthen its financial management by improving internal control systems and by implementing electronic systems that permit real-time reporting. These systems will allow the Department to optimize the use of Recovery Act funding in the creation of jobs and economic opportunities across the country.

As part of improving financial management and resource use, USDA will also build upon its performance measurement system to achieve further results. The Department will expand the use of performance metrics to track areas of success and those needing improvement. This information will allow agency decision makers to align resources to achieve the highest outcome. The Department also will use performance management strategies, including project labor agreements, to strengthen its contracting and procurement activities across the country.

USDA's plans include:

- Using such technology and process improvements as the Financial Management Modernization and the Web-based Supply Chain Management Initiatives to streamline operations;

- Promoting sound financial management through leadership, policy, and oversight;

- Using project labor agreements to ensure proper employment standards where feasible in contracting;

- Implementing strong and integrated internal control systems;

- Increasing the use of performance measurements and standards;

- Optimizing the use of American Recovery and Reinvestment Act resources;

- Implementing and maintaining an infrastructure to provide management with the real-time financial management information necessary for sound decision making;

- Improving cross-servicing for financial and administrative services to itself and other Federal Government agencies;

- Eliminating improper payments;

- Reviewing and addressing problem areas perceived to be affecting management efficiency; and

- Ensuring agency-developed material requiring action by the Office of the Secretary is analytically sound and consistent with Administration policy.

## Initiative V: Optimize Information Technology (IT) Policy and Applications

USDA is committed to increasing the economic opportunity and growth of rural communities across America. The Department is working to improve the effective delivery of programs and services to its constituents, applicants, and customers. The Department is deploying broadband, creating an enterprise platform that enables open communication channels, ensuring the protection of mission-critical operations and customer data, and supporting portfolio views for managing across organization and geographic boundaries. USDA plans to prioritize key technology investments through a re-engineering of the IT Capital Planning and Investment Control process. The Department also looks to ensure open, transparent, and collaborative avenues for easy access to USDA information. In addition, USDA aims to

protect the privacy of information collected in service delivery, and modernize foundational elements to consolidate and streamline core IT processes.

USDA's plans include:

- Prioritizing and optimizing all Departmental IT policies, programs, and spending;

- Leveraging resources and internal capabilities to reduce dependence on IT contractors;

- Using technology to improve program delivery and communication, including various IT modernization efforts, geospatial technologies, and social media;

- Integrating information systems to address such management challenges as the Comprehensive Information Management System (CIMS); and

- Implementing a modern, secure, and robust delivery platform across its enterprise.

## Initiative VI: Optimize USDA "Green" or Sustainable Operations

One of the President's top priorities for all Federal agencies is to establish an integrated strategy to work towards sustainability and reduce greenhouse gas emissions. As a steward of natural resources, USDA is committed to achieving these goals. The Department will focus its efforts towards sustainable operations. USDA aims to accomplish this task by: decreasing energy intensity; increasing renewable energy use; conserving water; promoting pollution prevention, waste reduction, and recycling; implementing sustainable building design, construction, and operation; increasing green procurement; promoting electronic product stewardship; and embracing environmental management systems to achieve sustainable operation goals.

USDA's plans include:

- Establishing aggressive, Departmentwide greenhouse gas emission goals;

- Monitoring greenhouse gas emissions through annual inventories; and

- Developing and implementing a Strategic Sustainability Performance Plan in accordance with Executive Order 13514.

## Initiative VII: Enhance USDA Homeland Security and Emergency Preparedness to Protect USDA Employees and the Public

USDA is working to enhance homeland security and emergency activities to provide a coordinated national effort to protect American agriculture and rural communities from intentional harm. The Department will ensure the Nation's quality of life through the continuance of a secure and reliable food supply. USDA will lead these efforts by protecting the food supply, maintaining security of USDA resources, securing infrastructure, and supporting emergency response and program operations nationwide.

USDA's plans include:

- Improving information technology security;
- Strengthening food and agriculture defense;
- Enhancing pandemic flu planning and response;
- Enhancing continuity plans; and
- Upgrading physical security at USDA facilities.

## Initiative VIII: Enhance the USDA Human Resources Process to Recruit and Hire Skilled, Diverse Individuals to Meet the Program Needs of USDA

USDA is reforming its hiring process to ensure a streamlined, user-friendly environment for both the applicant and the hiring manager. Such an environment will lead to the identification and selection of the most talented and competent workforce possible. In doing so, the Department will experience increased diversity while addressing current and future skills gaps. USDA is evaluating its human resources policies relative to talent management. Where necessary, the Department is also realigning its policies to further support its transformation of recruitment and retention initiatives. Additionally, USDA is addressing the gap between employee engagement and performance expectations.

USDA's plans include:

- Addressing current or future gaps in skill sets and workforce capacity;
- Increasing diversity in its workforce;
- Aligning its human resources policies;
- Streamlining hiring processes; and
- Establishing an employee satisfaction action team.

## Initiative IX: Enhance Collaboration and Coordination on Critical Issues through Cross-cutting Departmentwide initiatives

More than ever, the problems facing our customers require a holistic response. To enable agencies and programs to more effectively and efficiently achieve the strategic goals established in this plan, USDA will utilize cross-cutting initiatives to focus on the most critical and complex challenges. Initiatives do not perform programmatic activities; rather, they enhance the work already being done by USDA by offering an innovative environment for learning, sharing, and problem solving across traditional organizational boundaries. Examples of current collaborative, cross-cutting initiatives include the Biotechnology Working Group; *Know Your Farmer, Know Your Food*; and *Let's Move!*.

USDA's plans include:

- Establishing cross-cutting initiatives to more effectively address critical challenges;
- Identifying opportunities for collaboration across agencies; and
- Creating new, results-based reporting mechanisms to improve communication, problem solving, and decision making.

# Appendix A:
# Program Evaluations

USDA used several tools in developing this strategic plan, including:

- Program evaluations;
- Advisory committees;
- Office of Inspector General (OIG), Government Accountability Office (GAO), and other external reviews; and
- Internal management studies and performance measurement systems.

The following table highlights some of these tools as they relate to USDA's strategic goals.

| | Program Evaluations Used to Develop the Strategic Plan | | | |
|---|---|---|---|---|
| Goal | Evaluations/ Analyses | Brief Description | What Was The Effect | Date |
| Goal 1 | Independent reviews of Rural Development (RD) Customer Satisfaction Surveys | JD Powers review of RD performance when compared to mortgage companies proving single family home services | RD was rated "superior" | Every 2 years (most recently FY 2009) |
| | Comprehensive analysis of Food Safety and Inspection Service (FSIS) training programs | A meta-analysis of past evaluations of FSIS training programs | Findings from the contracted evaluation were confirmed and improvements to training programs were planned. These will help meet the outreach and training objectives of Goal 1 | October 2008 |
| | Economic Research Service (ERS) annual macroeconomic estimates | Review, analysis, and update of macroeconomic models that estimate number of jobs created and additional economic activity generated from the export of agricultural products at both farm and non-farm levels | Estimates the impact of agricultural export activity on jobs and income at both farm and non-farm levels | Annually |
| | GAO Beginning Farmers and Ranchers report | Report focused on coordination of USDA activities affecting beginning farmers and ranchers (all USDA agencies evaluated) | Congress provided mandatory funding for the Beginning Farmers and Ranchers Development Program | 2007 |
| | National Animal Health Laboratory Network (NAHLN) evaluation | The evaluation by NAHLN partners and stakeholders assessed how well the program meets its original objectives, future objectives, and what objectives need to change | Change in structure of NAHLN leadership, clearer definition of responsibilities, and initiation of closer examination of NAHLN progress and priorities | 2007 |
| | American Customer Satisfaction Index comprehensive program evaluation for the Farmland Ranch Lands Protection Program (FRPP) | Program delivery assessment for program administered by Natural Resources Conservation Service (NRCS) | All Federal Government programs have a combined average of 69 for an ACSI indexing score; FRPP received a 73 | Completion and final report available as of September 2009 |

| | Program Evaluations Used to Develop the Strategic Plan | | | |
| Goal | Evaluations/ Analyses | Brief Description | What Was The Effect | Date |
|---|---|---|---|---|
| Goal 1 (cont.) | Oversight of Recovery Act monies expended by USDA programs | Oversight of both funded and unfunded Emergency Watershed Protection Floodplain Easement applications | NRCS developed a national oversight and evaluation plan | September 2009 |
| Goal 2 | Asian Longhorned Beetle Eradication Program review | Review to determine if eradication of this devastating forest pest is still possible, given recent detections and available resources | Final report pending | Ongoing |
| Goal 3 | Foreign Agricultural Service (FAS) Survey of Non-Governmental Organizations (NGOs) | Survey of NGOs who distribute school lunches and other assistance available through the McGovern-Dole Food for Education and Child Nutrition Program | Assists FAS program managers and NGOs in better targeting program resources | Ongoing |
| | Joint Command Provincial Reconstruction Team Evaluations | Periodic progress reviews of U.S. Governmentwide effort to rebuild war-torn Afghan provinces | Provides feedback on progress made to assist Afghan provinces in attaining and sustaining food security and self-sufficiency | Ongoing |
| | ERS Food Aid Targeting Effectiveness Measure | Review, analysis, and update of selected criteria, and running of an econometric model for selected food-insecure countries | Provides feedback to program managers to assist them in targeting food-insecure populations throughout the world | Annually |
| | Food Defense Plan Survey | Gathers data about industry's voluntary adoption of food defense plans | Survey results indicate an increasing trend in the number of establishments adopting a Food Defense Plan | Surveys: August 2006, November 2007, August 2008 |
| | ERS report on Household Food Security in the United States, 2007 | Provides national estimates of the prevalence of food security, food insecurity, and very low food security | Serves as a broad outcome measure for the effectiveness of nutrition assistance and provides metric for President's goal to end childhood hunger | Annually (most recently November 2008) |
| Goal 4 | Reports on Supplemental Nutrition Assistance Program (SNAP) Participation Rates, 2000 to 2007 | Reports national rates of participation among eligible people | Showed that while SNAP is reaching the neediest eligible individuals, many eligible people are not participating | June 2009 |
| | Access, Participation, Eligibility and Certification (APEC) study for school meals programs | Provides an estimate of erroneous payments in the school meals programs | Showed that errors in certification, counting, and claiming are significant sources of improper payments | November 2007 |
| | Assessment of program impacts on diet quality | Uses data from the National Health and Nutrition Examination Survey | Showed that, while programs are linked to certain positive dietary attributes, the diets of all groups (low-income participants, nonparticipants and higher-income individuals) fell far short of the Dietary Guidelines for Americans | July 2008 |
| | Review of the Use of Process Control Indicators in the FSIS Public Health Risk-Based Inspection System | Letter Report by the National Academy of Sciences (NAS) | FSIS revised its technical report for the public health decision criteria and began using its public Health Decision Criteria in July 2009 | October 2009 |
| | Review of the Food Safety and Inspection Service Risk-Based Approach to Public Health Attribution | Letter Report by NAS | FSIS continues to develop methods for including serotype information in its attribution work | October 2009 |

| Program Evaluations Used to Develop the Strategic Plan | | | | |
|---|---|---|---|---|
| Goal | Evaluations/ Analyses | Brief Description | What Was The Effect | Date |
| All Goals | National Institute of Food and Agriculture (NIFA) portfolio assessments | Assessments focus on current and/or emerging issues of societal importance | Improvements in program-planning and management, informing performance-based budgeting and staffing changes | Annually |
| | Improper Payments Information Act (IPIA) reviews | Analysis and testing using required IPIA thresholds | Reduction in amount of improper payments | Annually |
| | OIG and GAO audits and reviews | Financial statement audited annually and programs reviewed on a variable schedule | Staff made improvements to address recommendations | Annually |

USDA will undertake many new and ongoing evaluations over the next 5 years. The following table highlights some of the longer-term studies as they relate to USDA's strategic goals.

| Future Program Evaluations and Other Analyses | | | | |
|---|---|---|---|---|
| Goal | Evaluations/ Analyses | General Scope | Methodology | Timetable |
| Goal 1 | OIG review of National Organic Program | Audit to evaluate whether agricultural products marketed as organic meet the requirements of the program and to assess the adequacy and consistency of program oversight | OIG will issue a report and USDA's Agricultural Marketing Service will respond to the recommendations as appropriate | FY 2010 |
| | National Rabies Management Program review | Conduct a review in two phases: Phase I – goals, timelines, management strategies, and benefits vs. costs; and Phase II – administrative processes, budget, supervisory controls, internal and external networks, and communications | An internal Animal and Plant Health Inspection Service (APHIS) review team will provide a report of findings and recommendations for adjusting policies, practices, and processes as they relate to overall program objectives | FY 2010 |
| | Packers and Stockyards Program Management Accountability Review | Conduct internal management accountability reviews of all major Packers and Stockyard Program units to measure performance and ensure conformance with established standard operating procedures | The Grain Inspection, Packers and Stockyards Administration (GIPSA) will perform independent audits of each major organizational unit, based on established auditing procedures and business criteria. The program will hire an external entity to facilitate the reviews. | FY 2010 |

| | Future Program Evaluations and Other Analyses | | | |
|---|---|---|---|---|
| Goal | Evaluations/ Analyses | General Scope | Methodology | Timetable |
| Goal 1 (cont.) | Economic Research Service (ERS) annual macroeconomic estimates | Review, analyze, and estimate the impact of agricultural export activity on jobs and income at both farm and non-farm levels. | ERS uses sophisticated econometric modeling techniques to estimate the macroeconomic impact of exports on various economic criteria, including jobs and income, at both farm and non-farm levels. | Annually |
| | Analysis of the Federal Crop Insurance Corporation's (FCIC) product portfolio | Comprehensive review of the risk-management products offered by FCIC | Actuarial and underwriting experts review current and proposed crop insurance products and opportunities to assist the FCIC Board in developing a product strategy | Ongoing |
| | Review of FCIC policies, plans of insurance, and related materials | Comprehensive quality review of FCIC's materials | Actuarial and underwriting experts review FCIC legislation, regulation, and program materials to recommend potential ways to improve the overall quality of the program | Ongoing |
| | RD Customer Satisfaction Surveys | Independent reviews of Customer Satisfaction Evaluations | Surveys are planned for Water and Environmental Programs Direct Loans and Grants, Business & Industry Guaranteed Loans, and Multi-family Housing Guaranteed Loans | Every 2 years |
| Goal 2 | Conservation Effects Assessment Project (CEAP) | Conservation Effects Assessment Project (CEAP) | CEAP research and literature surveys and reports informed development of performance measures. | FY 2010 (CEAP Upper Mississippi River Basin report) |
| | Evaluation of Watersheds Program | Comprehensive cost benefits analysis of watersheds program | Pilot for developing a process for recurring, comprehensive programmatic evaluations | May-September 2010 |
| | Asian Longhorned Beetle Eradication Program strategy review | Review of the current program protocols employed by the Asian Longhorned Beetle Eradication Program | Combine Asian longhorned beetle program operational knowledge with expert understanding of beetle biology to determine how protocols can be revised to achieve the stated objectives | FY 2010 |
| Goal 3 | Biotechnology Petition Process Improvement Initiative | Review, analysis, and recommendation of improvements for the regulatory determination process related to biotechnology product petitions | Internal evaluation group will review and report with recommendations to be considered for incorporation into the process | FY 2010 |
| | Assessment of functionality and usability of the Biotechnology Regulatory Services Web site | Analysis of content, layout, organization, usability, and functionality of the Web site for customers, stakeholders, and the public | Contractor will be hired to conduct focus groups with customers and benchmark with the Websites of other Federal agencies to identify best practices | FY 2010 |
| | FAS survey of non-governmental organizations (NGOs) | NGOs that distribute school lunches and other assistance available through the McGovern-Dole Food for Education and Child Nutrition Program | Standardized survey and reporting tool used by NGOs to report results of program outlays in various countries targeted for program resources | Ongoing |

| Future Program Evaluations and Other Analyses | | | | |
|---|---|---|---|---|
| Goal | Evaluations/ Analyses | General Scope | Methodology | Timetable |
| Goal 3 (cont.) | ERS Food Aid Targeted Effectiveness measure | Estimate how effectively the U.S. Government is targeting its program resources aimed at reducing world hunger | ERS uses sophisticated econometric modeling techniques to estimate the effectiveness of U.S. Government food aid distribution through its various program resources | Annually |
| | Biotech Petition Process Improvement Initiative | Review, analyze, and recommend improvements for regulatory determination process | Internal agency evaluation group will conduct review and provide report with recommendations | FY 2010 |
| | Joint Command Provincial Reconstruction Team Evaluations | Review, analyze, and make adjustments as necessary to maximize effectiveness of efforts to rebuild basic Afghan infrastructure | Provincial Reconstruction Teams are multi-disciplinary and meet regularly to coordinate massive effort to help Afghan provinces attain food security | Ongoing |
| Goal 4 | Sterile fruit fly production review | Review the current status of sterile fruit fly production | International team of fruit fly experts will conduct a review and provide a final report with recommendations | FY 2010 |
| | Citrus Health Response program review | Evaluate overall program strategies for controlling Asian citrus psyllids and the regulatory framework to prevent the spread of citrus greening | Internal APHIS review team will provide a report including recommendations as they relate to overall program objectives | FY 2011 |
| | Evaluate effectiveness of foodborne illness outbreak investigations | Conducted in response to the President's Food Safety Working Group recommendations | Document reviews, interviews, and quantitative data analysis | March 2011 |
| | Analyze food safety assessment results from official establishment subject to testing for E. coli O157:H7 | Conducted to determine if changes are needed to food safety assessment methods or policy to better prevent product adulteration by E. coli O157:H7 | Document reviews and quantitative data analysis, including statistical analysis | FY 2010 |
| | Data Coordination Committee (DCC) | Identify data collection, quality, or analysis issues for discussion at DCC meetings. | Assess new research and ensure that projects are harmonized across FSIS | Bimonthly |
| | Development of Data Analysis Sections for FSIS Notices and Directives | Identify data analyses that should be performed in response to FSIS Notices and Directives | Implement applicable data analyses in each notice/directive | As needed |
| | Food Defense Plan Survey | Gathers data about industry's voluntary adoption of food defense plans | FSIS Inspection Program Personnel will receive the survey questions through the Performance-Based Inspection System | FY 2010 |
| | Measures of Nutrition Assistance Program Coverage | Estimates the rate of participation among eligible people for SNAP and WIC | Compares econometric models of the eligible population based on demographic and economic data to administrative data on participation | Annual for SNAP; periodic for other programs |

| | | Future Program Evaluations and Other Analyses | | |
|---|---|---|---|---|
| Goal | Evaluations/ Analyses | General Scope | Methodology | Timetable |
| Goal 4 (cont.) | Erroneous Payment Measures | Estimates of erroneous payments from major Federal nutrition assistance programs | Varies by program; generally involves analysis of program operations data supplemented by special data collections on recipient/program delivery partner characteristics | Every 5-10 years (varies by program), with interim indicator measures on erroneous payment risks |
| | Evaluation of Fresh Fruit & Vegetable Program (FFVP) | Determine whether participation in the FFVP increases children's consumption of fruits and vegetables | Compare experiences of students in FFVP schools with similarly situated non-participating schools, using 24-hour dietary recalls, Web-based surveys, and interviews with school officials, teachers, students, and parents | Findings in late 2012 |
| | School Nutrition Dietary Assessment Study-IV | Assesses the nutrient content of school meals, the school environment that affects the programs, and other aspects of Federal school meal programs | Data will be collected from a nationally representative sample of districts and schools in school year 2009-2010 | Findings in 2011 |
| All Goals | NIFA – Portfolio Assessments | Assessing NIFA portfolios of programs on relevance, quality, and performance | External expert review with self-assessments occurring in intervening years | Ongoing |

# Appendix B:
# Cross-Cutting Programs

USDA's work often cuts across jurisdictional lines within USDA, with other Federal agencies, and with State, local, and private partners. This table lists the primary partnerships that will enable USDA to reach the outcomes in this Strategic Plan. Please note that for the purposes of this table, it is assumed that all USDA Departmental Offices support all strategic goals and management initiatives.

| Cross-cutting Programs | | |
|---|---|---|
| **Goal** | **USDA Primary Agencies** | **External Organizations** |
| Goal 1 | Agricultural Marketing Service (AMS), Animal & Plant Health Inspection Service (APHIS), Economic Research Service (ERS), Farm Service Agency (FSA), Food Safety Inspection Service (FSIS), Foreign Agricultural Service (FAS), Forest Service (FS), Grain Inspection, Packers & Stockyards Administration (GIPSA), National Agricultural Statistics Service (NASS), National Institute of Food and Agriculture (NIFA), Natural Resources Conservation Service (NRCS), Office of Budget and Program Analysis (OBPA), Office of the Chief Economist (OCE), Office of the Chief Financial Officer (OCFO), Office of Civil Rights, Office of the Inspector General (OIG), Risk Management Agency (RMA), Rural Development (RD), World Agricultural Outlook Board (WAOB) | Agribusiness industry officials, America On the Move Foundation, American Bar Association, American Farm Bureau Federation, American Savings Education Council, Bio-Based Products and Bio-Energy Coordination Council, Centers for Disease Control and Prevention (CDC), commercial lenders, commercial warehouse operators, Commodity Futures Trading Commission, commodity groups (cooperators), Department of Commerce, Department of Defense (DOD), Department of Education, Department of Energy (DOE), Department of Health and Human Services (HHS), Department of Homeland Security (DHS), Department of Housing and Urban Development, Department of the Interior (DOI), Department of Transportation, Environmental Protection Agency (EPA), Farm Foundation, Federal Emergency Management Agency (FEMA), Financial Literacy and Education Commission, Food and Drug Administration (FDA), Helping America's Youth, Internal Revenue Service, Jump$tart Coalition for Personal Financial Literacy, land-grant and other universities and colleges, National Aeronautics and Space Administration (NASA), National Animal Rescue and Sheltering Coalition, National Institutes of Health, National Oceanic and Atmospheric Administration (NOAA), National Savings Forum, National Science Foundation (NSF), Office of U.S. Trade Representative, private and cooperative lending institutions, private industry trade groups, private sector insurance companies, producers, regional development banks, Small Business Administration, State Agriculture Finance Programs, State and regional trade associations, State departments of agriculture, U.S. Army, U.S. Agency for International Development, U.S. Geological Survey (USGS), and Veterans Administration |
| Goal 2 | Agricultural Research Service (ARS), Center for Nutrition Policy and Promotion (CNPP), ERS, Food and Nutrition Service (FNS), FS, FSA, NASS, NIFA, and NRCS | Bureau of Land Management, Bureau of Reclamation, conservation groups (e.g., Ducks Unlimited, Pheasants Forever), DOI – Fish and Wildlife Service, EPA, FEMA, land-grant and other universities and colleges, National Association of Conservation Districts, National Park Service (NPS), NOAA, regional air quality planning organizations, research partnerships (with universities, USGS, non-governmental organizations (NGOs), etc.), State agencies, State soil and water conservation districts, tribal governments, U.S. Army Corps of Engineers, and USGS |
| Goal 3 | APHIS, ARS, FAS, and Research, Education and Economics (REE) | Agribusiness industry officials, commodity groups (cooperators), Department of Commerce, Department of State, DOD, educational and research institutions, FDA, International Trade Commission, land-grant and other universities and colleges, NGOs, Office of U.S. Trade Representative, regional development banks, State and regional trade associations, United Nations World Food Programs, U.S. Agency for International Development (USAID), The White House, and World Trade Organization |
| Goal 4 | AMS, APHIS, CNPP, FAS, FNS, FS, FSA, NIFA, FSIS, and WAOB | American Association of Veterinary Diagnostic Laboratory Diagnosticians, American Nursery and Landscape Association, American Peanut Shellers Association, Association of Food and Drug Officials, Association of Home Appliance Manufacturers, commodity livestock organizations including livestock markets, allied industries, academia, and State brand inspection systems, CDC, Department of Justice, Department of Commerce, Department of State, DHS, DOD, DOE, DOI, Electronic Warehouse Receipt Providers, EPA, extension agents, Federal Bureau of Investigation, FDA - Incident Command System, Foodborne Illness Attribution Workgroup, FoodNet, heath and public interest organizations, HHS, HHS - National Biosurveillance Integration System, International Food Information Council, Joint Institute for Food Safety and Applied Nutrition, |

| Cross-cutting Programs | | |
| Goal | USDA Primary Agencies | External Organizations |
| --- | --- | --- |
| Goal 4 (cont.) | | land-grant universities, National Assembly of State Animal Health Officials, National Association of State Departments of Agriculture, National Association of State Meat and Food Inspection Directors, National Cotton Council, National Grain and Feed Association, National Institute of Food and Agriculture, National Plant Board, National Plant Diagnostic Network, NSF, National Science Teachers Association, North American Millers' Association, Partnership for Food Safety Education, private non-profit voluntary organizations, private sector firms and organizations, private voluntary organizations, professional organizations, Society of American Florists, Specialty Crop Farm Bill Alliance, State and local health departments, State and university veterinary diagnostic laboratories, State, territorial, tribal, and local agencies involved in nutrition assistance program delivery, tribal governments, USAID, U.S. Animal Health Association, and U.S. Customs and Border Protection |

# Appendix C:
# Strategic Consultations

USDA regularly consults with external stakeholders, including USDA's customers, partners, landowners, policy experts, and industry and consumer groups regarding our programs' effectiveness. While many of the consultations were not conducted expressly for the development of USDA's Strategic Plan, they did impact strategic goals, objectives, strategies, and targets. Additionally, the Strategic Plan was developed in accordance with guidance from the Office of Management and Budget (OMB) and the Government Performance and Results Act (GPRA).

With the full support of its senior leadership, USDA regularly consults with stakeholders and seeks validation of all goals, objectives and performance measures from employees and the public.

| Strategic Consultations | | | |
|---|---|---|---|
| Goal | Date | Who | Purpose |
| All Goals | May – October, 2009 | Secretary Vilsack hosted 22 listening sessions or "Rural Forums" with residents in small communities across the country as part of President Obama's "Rural Tour." | Listening to diverse voices in communities throughout rural America and learning from citizens what the Administration can do to strengthen the rural communities. Frequently addressed topics included: jobs and quality-of-life issues, climate change and energy legislation, rural infrastructure, and concerns about the dairy industry. |
| Goal 1 | Ongoing | Commodity Futures Trading Commission | Monitoring and surveillance of prices in the spot and futures markets for livestock to ensure competitive markets |
| | December 8-10, 2009 | Federal Emergency Management Agency, Department of Health and Human Services, animal humane organizations, and States | Develop and refine objectives and strategies for effective disaster response efforts for companion animals |
| | Ongoing through FY 2010 | Department of Justice | Joint public workshops exploring competition issues affecting the agriculture industry in the 21st century and the appropriate role for antitrust and regulatory enforcement in that industry |
| | Ongoing | Producers and producer groups | Discussing any proposed new crop insurance program or evaluations of existing programs |
| | Ongoing | Crop insurance companies | Discussing operations and plan and implement program improvements |
| | Ongoing | University risk management specialists and economists | Review proposed new crop insurance tools or programs; conduct research and education |
| | Ongoing | State and local governments and other Federal agencies | Coordinate risk-management initiatives |
| | Ongoing | Producers, producer groups and associations, commercial lenders, and land-grant colleges and universities | Share information and provide input on program delivery and outreach |
| | Ongoing | AARP Foundation, American Council on Consumer Interests, American Humane Association, Association for Financial Counseling and Planning Education, Consumer Federation of America, Council for Agricultural Science and Technology, Federal interagency working groups and task forces, industry groups, land-grant and other universities and colleges, National Association of Counties, National Association of Elementary School Principals, National Endowment for Financial Education, National Home Safety Council, National Institute for American Agriculture, other Federal agencies, practitioner associations, producer groups, and science societies and associations | Stakeholder input and partner coordination for program planning |

| | | Strategic Consultations | |
|---|---|---|---|
| Goal | Date | Who | Purpose |
| Goal 1 (cont.) | Annually (most recently November 2009) | U.S. Agricultural Export Development Council participants | Annual conference that brings together U.S. Government officials, agribusiness industry groups, and State and regional trade associations to work in partnership with USDA in market-development and export-promotion programs. Purpose is to share information and best practices, with the goal of improving coordination and effectiveness of U.S. agricultural export promotion efforts. |
| | November 2009 | Listening Sessions with tribal leaders and representatives | Dialogue conducted in conjunction with the President's Tribal Leaders Conference |
| Goal 2 | Ongoing | Federal interagency working groups and task forces, land-grant and other universities and colleges, National Council for Science and the Environment, other Federal agencies, and science societies and associations | Stakeholder input and partner coordination for program planning |
| | August 13, 2009 | National Plant Board | Dialogue with National Plant Board members about strategies to address pest risks via the firewood pathway and to develop a comprehensive multiparty strategy for firewood |
| | November 9, 2009 | Massachusetts Department of Conservation and Recreation, Massachusetts Department of Agricultural Resources, U.S. and State legislators, City of Worcester, and Towns of Holden, Shrewsbury, Boylston, and West Boylston | Presented the strategic plan for fiscal year (FY) 2010 to those impacted by the Asian Longhorned Beetle Program in Worcester County, Massachusetts, and revised plans based on feedback from those present |
| | Ongoing | Public | Obtained public comments on interim final rules and proposed rules for the 2008 Farm Bill programs. Natural Resources Conservation Service (NRCS) is in the process of developing final rules, giving full consideration to the public comments received. |
| | July 14, 2008 | Federal, State, and local government agencies, academic professionals, non-government organization (NGO) representatives, and the public | Listening session on the Chesapeake Bay Watershed Initiative authorized under the 2008 Farm Bill. More than 250 stakeholders provided ideas about priority practices, programs, and geographic areas. |
| | Ongoing | Conservation Districts, Resource Conservation and Development Program Councils, State conservation agency partners, NRCS field representatives, and the public | Listening sessions and focus groups for the Soil and Water Resources Conservation Act to obtain partner and public input on priority natural resource concerns and potential approaches to strengthening conservation adoption |
| Goal 3 | Annually (most recently April 2009) | International Food Aid Conference participants | Annual conference jointly hosted by USDA and U.S. Agency for International Development (USAID) brings together individuals involved in food-assistance activities, across the United States and throughout the world. Aim is to share information on best practices and improve coordination and effectiveness of food-assistance activities in various countries and regions of the world. |
| | April 29-30, 2009 | Stakeholders and the public | Issue-focused public meeting to solicit feedback on revisions to existing regulations regarding the importation, interstate movement, and environmental release of certain genetically engineered organisms |
| | Ongoing | Department of Commerce, Department of Defense, Department of State, USAID, multinational organizations, and educational and research institutions | Iraq and Afghan Provincial Reconstruction and Development. These war-torn countries require vast resources to rebuild and develop basic infrastructure to enable them to have the capacity to feed themselves throughout the year. Because of the agrarian nature of these populations and their unique needs, USDA is a major participant in these ongoing efforts, which require constant joint strategic consultations involving multidisciplinary teams with the expertise to facilitate these efforts. |

| Strategic Consultations | | | |
|---|---|---|---|
| Goal | Date | Who | Purpose |
| Goal 4 | Quarterly | State animal health officials and laboratory directors | Strategic planning and implementation of veterinary diagnostic testing under the National Animal Health Laboratory Network to increase capabilities and capacities for early detection, rapid response, and appropriate recovery from animal health emergencies |
| | June 17-18, 2009 | Department of Homeland Security's Customs and Border Protection, State officials and industry representatives from the U.S., Mexico, and Belize | Reviewed citrus health status in the U.S. and Mexico regarding Citrus Greening and Asian Citrus Psyllid and developed a tri-national approach |
| | June 2010 | State departments of agriculture and specialty crop industry representatives | Annual Farm Bill Section 10201 meeting to gather stakeholder input for activities to be conducted in FY 2011 |
| | Biennially | State departments of agriculture and specialty crop industry representatives | Biennial meeting of the Cooperative Agricultural Pest Survey community and state cooperators to obtain program input on the future direction of the Cooperative Agricultural Pest Survey program |
| | Ongoing | State and local partners and the public | USDA's Food and Nutrition Service hosts listening sessions, including webinars and other Web-based communication strategies and seeks public input to better understand the needs and concerns of program participants and partners in preparation for reauthorization of programs |
| | Ongoing since October 2008 | 2010 Dietary Guidelines Advisory Committee | Review *Dietary Guidelines for Americans* in light of current scientific and medical knowledge; recommend changes |
| | Ongoing | America On the Move Foundation, American Community Gardening Association, Association for Size Diversity and Health, Community Food Security Coalition, Federal interagency working groups and task forces, industry groups, land-grant and other universities and colleges, other Federal agencies, and science societies and associations | Stakeholder input and partner coordination for program planning |
| | Ongoing since October 2009 | National Academy of Sciences | Peer review projects and initiatives proposed by the Department to ensure that USDA decisions are science-based and data-driven |
| | Ongoing | Outside peer review | Food Safety and Inspection Service (FSIS) utilizes outside agencies and consulting firms to provide outside review of FSIS projects and initiatives, such as the Foodborne Illness Attribution work currently being conducted. |
| | Ongoing | National Advisory Committee on Microbiological Criteria for Foods | Providing impartial, scientific advice to Federal food-safety agencies for use in the development of an integrated national food safety system approach from farm to final consumption, to ensure the safety of domestic, imported, and exported foods |
| | Ongoing | National Advisory Committee on Meat and Poultry Inspection | Advising the Secretary of Agriculture on matters affecting Federal and State inspection program activities |
| | Ongoing | Safe Food Coalition | Monthly meeting with the consumer group to discuss food safety issues |
| | Ongoing | Industry representatives | Monthly meeting to discuss food safety and regulatory issues |
| | Ongoing | Food Safety Working Group (FSWG) | Through collaborative partnerships with consumers, industry, and regulatory partners, the FSWG is committed to modernizing food safety, with a public health-focused approach to food safety based on three core principles: (1) prioritizing prevention; (2) strengthening surveillance and enforcement; and (3) improving response and recovery. |

| Strategic Consultations | | | |
|---|---|---|---|
| Goal | Date | Who | Purpose |
| Goal 4 (cont.) | Ongoing | Partnership for Food Safety Education | The Partnership for Food Safety Education unites industry associations; professional societies in food science, nutrition and health; consumer groups; and the U.S. Government to educate the public about safe food handling. |

THIS PAGE INTENTIONALLY LEFT BLANK

www.ingramcontent.com/pod-product-compliance
Lightning Source LLC
Chambersburg PA
CBHW081243180526

45171CB00005B/522

*9781479318940*